Basics of communications and coding

Basics of communications and coding

William G. Chambers
Department of Mathematics,
Westfield College,
University of London

CLARENDON PRESS · OXFORD
1985

Oxford University Press, Walton Street, Oxford OX2 6DP
London New York Toronto
Delhi Bombay Calcutta Madras Karachi
Kuala Lumpur Singapore Hong Kong Tokyo
Nairobi Dar es Salaam Cape Town
Melbourne Auckland
and associated companies in
Beirut Berlin Ibadan Mexico City Nicosia

Oxford is a trade mark of Oxford University Press

Published in the United States
by Oxford University Press, New York

© William G. Chambers, 1985

All rights reserved. No part of this publication may be reproduced, stored in a retrieval system, or transmitted, in any form or by any means, electronic, mechanical, photocopying, recording, or otherwise, without the prior permission of Oxford University Press

This book is sold subject to the condition that it shall not, by way of trade or otherwise, be lent, re-sold, hired out or otherwise circulated without the publisher's prior consent in any form of binding or cover other than that in which it is published and without a similar condition including this condition being imposed on the subsequent purchaser

British Library Cataloguing in Publication Data
Chambers, William G.
   Basics of communications and coding.
   1. Digital communications   2. Coding theory
   I. Title
   621.38'0413   TK5103
   ISBN 0-19-853195-8
   ISBN 0-19-853194-X Pbk

Library of Congress Cataloging in Publication Data
Chambers, William G.
   Basics of communications and coding.
   Bibliography: p.
   Includes index.
   1. Digital communications.   2. Coding theory.
3. Statistical communication theory.   I. Title.
TK5103.7.C53   1985   621.38'0413   84-23006
ISBN 0-19-853195-8
ISBN 0-19-853194-X (pbk).

Set by Joshua Associates, Oxford
Printed in Great Britain by
St. Edmundsbury Press Ltd,
Bury St. Edmunds, Suffolk.

To Bernardine

Preface

This book grew out of a course taught at the University of London, intended for third-year and graduate students of electronics, computer science, mathematics, and physics. It thus attempts to reach a wider audience than a traditional text on communications. None the less it covers what is normally regarded as the basic material of traditional communications theory, although not always in such detail as a typical engineering text.

In keeping with modern trends the emphasis is totally on digital communications. Analogue communication systems such as FM radio are not discussed since (a) they are well-covered in the classical texts, (b) they are of decreasing importance in comparison with digital systems, (c) the concepts do not seem to be so interesting as those of digital communications, so that the limited space available is better used on the latter. None the less much of the modulation theory described can easily be adapted for dealing with analogue signalling.

The book starts in a fairly traditional manner, with chapters on Fourier theory, digital signalling methods, probability theory and noisy channels. However it then goes on to cover several topics not usually covered at this level. (a) There is a fairly thorough discussion of Shannon's revolutionary discovery that with suitable coding information transmission over a noisy channel at a given rate not set too high can be carried out with a probability of error as small as one pleases. This is first illustrated using a simple form of modulation known as pulse-position modulation. Shannon's theorem is then discussed as it applies to the band-limited Gaussian channel. (b) Error-correcting codes are introduced in the context of Gaussian channels as forms of modulation which lower the probabilities of error. (c) In an attempt to allay the idea that Shannon's work is rather remote from everyday life a family of powerful and useful error-correcting codes is introduced, capable of approaching the Shannon limit. (d) The advent of optical communications means that the limitation imposed by quantum theory on information rates is now an important topic. This topic is discussed from an elementary point of view, but without sacrificing accuracy of presentation. (e) Cipher systems are becoming increasingly important in communications systems, because of the risk of interception and tampering to which signals sent by radio are prone. Moreover there have been some exciting theoretical developments in this field in the last few years.

The topics have been selected so that the mathematics is both conceptually straightforward and also useful in other contexts. Three chapters cover the mathematical background. They have been spaced out so that applications of the techniques follow the introduction of the techniques as soon as possible. Of necessity the treatments are brief, so that some familiarity is helpful, but they are also intended to be as far as possible complete for the purpose in hand.

This is to minimize the number of gaps in the main lines of development needing to be filled by appeal to external authority or by the use of material in other texts. Of the mathematical topics, most third-year students will have some familiarity with Fourier transforms and probability theory. The third topic, the theory of finite fields, is not usually found in elementary texts on communications theory. However it seems that it is becoming as much part of the repertoire of communications theory as the other topics. Although coding has been introduced fairly early on as a form of digital modulation and as a natural extension of modulation techniques like signal-constellations, none the less the use of algebraic structures in coding has been left until later. Algebraic codes are then treated in reasonable detail. Fortunately the algebra used here closely mimics ordinary polynomial algebra, and so the reader is never very far from what is familiar to him.

It may be worth drawing the reader's attention to the index. A lot of trouble has gone into preparing it. If a technical term or acronym is used without definition or cross-reference, then the index should indicate where it is first used.

Acknowledgements

I would like to express my gratitude above all to Paddy Farrell for his advice and encouragement, and also to many friends, colleagues, and visitors to Westfield College for their help in the preparation of this work. In particular I would like to mention Fred Piper, Thomas Beth, Dwijendra Ray-Chaudhuri, Gary Gledhill, Terry Parker, and Peter Williams. I would also like to thank the students whose comments on the presentation have proved very helpful, and the staff of the Westfield computer unit for their assistance in the preparation of the manuscript and of the diagrams.

London
May 1984

W. G. C.

Contents

Glossary of function names and symbols		xv
1	**Introduction**	1
1.1	Aims and outline	1
	1.1.1 Introductory comments	1
	1.1.2 Communications	1
	1.1.3 The robustness of digital communications	2
	1.1.4 Reliable communications	3
	1.1.5 Outline	4
1.2	Physical background	6
	1.2.1 Frequency bands	6
	1.2.2 Power and power flux	8
	1.2.3 The decibel	9
	1.2.4 Noise and interference	10
	Exercises	11
2	**Fourier theory**	13
2.1	The Fourier transform	13
	2.1.1 The Fourier inversion theorem	13
	2.1.2 A few simple results	16
	2.1.3 Convolutions and correlations	17
2.2	The delta function	19
	2.2.1 Definition and properties	19
	2.2.2 The Fourier transform	20
	2.2.3 Filters	20
	2.2.4 Impulse response function	21
	2.2.5 Transfer function	22
	2.2.6 Some examples of filters	23
2.3	Periodic functions	26
	2.3.1 Fourier series	26
	2.3.2 A few simple results	29
	2.3.3 Filtering a periodic function	30
	2.3.4 Periodically repeated functions	32
	2.3.5 The looped wave-guide	32
2.4.	Delta combs	34
	2.4.1 Fourier transform of a periodic function	34
	2.4.2 The periodic comb	34
	Exercises	36

CONTENTS

3 **Digital signalling methods** 39

   3.1 The concept of information 40

      3.1.1 Information and information rates 40
      3.1.2 Data compression 41

   3.2. Baseband signalling 43

      3.2.1 Baseband binary and multilevel pulse signalling 43
      3.2.2 Information rate for 'moderate reliability' 44
      3.2.3 Gray codes 45
      3.2.4 Codes and signals 45
      3.2.5 Nyquist sampling of analogue signals 47
      3.2.6 Analogue-to-digital conversion 49
      3.2.7 Links using repeaters 51

   3.3 Band-restricted signalling 52

      3.3.1 Band-restricted pulse signalling 52
      3.3.2 Intersymbol interference 56
      3.3.3 Adaptive equalizers 56
      3.3.4 Duobinary system 58

   3.4 Carrier-based signalling 60

      3.4.1 Carrier modulation 60
      3.4.2 Positive-frequency and phasor representations 61
      3.4.3 Quadrature modulation 63
      3.4.4 Carrier-based pulse signalling 63
      3.4.5 Frequency-shift keying 65

   Exercises 66

4 **Random variables and noise** 69

   4.1 Probability theory 69

      4.1.1 Basic concepts 69
      4.1.2 Random variables 72
      4.1.3 Continuous random variables 73
      4.1.4 The Chernoff bound 74
      4.1.5 Gaussian distributions 76
      4.1.6 Multivariate Gaussian distributions 77

   4.2 Noise signals 77

      4.2.1 Introductory 77
      4.2.2 The autocorrelation function 78
      4.2.3 The effect of filtering on noise 80
      4.2.4 White noise 80

CONTENTS xi

		4.2.5 Shot noise	82
		4.2.6 Gaussian noise	85
	4.3	Gaussian noise in polar signalling	86
		4.3.1 Introductory	86
		4.3.2 Integrating a single pulse	86
		4.3.3 Error probability	87
		4.3.4 Reliable communication	87
	4.4	Rayleigh fading	89
		4.4.1 Rayleigh channel	89
		4.4.2 Error probability	90
	Exercises		91

5	**Signal spaces**		94
	5.1	Expansions as function spaces	94
		5.1.1 Introductory	94
		5.1.2 Dimension and bandwidth equivalent	95
		5.1.3 Scalar product and length	96
		5.1.4 Expansions with an orthonormal basis	97
		5.1.5 Some examples of signal spaces	98
		5.1.6 Hadamard functions	99
		5.1.7 Matched filters	101
		5.1.8 The effect of additive white noise	102
	5.2	Codes in signal-space	103
		5.2.1 Introductory	103
		5.2.2 Lattice codes	106
		5.2.3 Spherical codes	107
		5.2.4 Orthogonal and biorthogonal signalling	107
	5.3	Shannon's theorem	111
		5.3.1 Introductory	111
		5.3.2 The theorem	111
		5.3.3 Random coding	114
		5.3.4 The converse theorem	114
		5.3.5 Reliability and complexity	114
		5.3.6 Wide-band and narrow-band signalling	115
		5.3.7 To code or not to code	116
	Exercises		119

6 Error-correcting codes — 122

6.1 Codes and channels — 122
- 6.1.1 The BSC and binary codes — 122
- 6.1.2 Feedback and forward error-correction — 125
- 6.1.3 Simple codes on the AWGN channel — 126
- 6.1.4 Burst channels — 127

6.2 Trellis codes and the Viterbi algorithm — 128
- 6.2.1 The duobinary code as a trellis code — 128
- 6.2.2 The Viterbi algorithm — 130
- 6.2.3 MSK modulation and multi-h phase codes — 130
- 6.2.4 Convolutional codes — 132

6.3 Binary linear block codes — 133
- 6.3.1 Introductory — 133
- 6.3.2 Binary hard-decision decoding — 134
- 6.3.3 The parity-check matrix and the syndrome — 137
- 6.3.4 Hamming codes — 138
- 6.3.5 Two particular codes — 139

6.4 Binary block codes on the AWGN channel — 140
- 6.4.1 Binary linear codes on the AWGN channel — 140
- 6.4.2 Performance of Reed–Muller codes — 141
- 6.4.3 Weakness of hard-decision decoding — 143
- 6.4.4 Hard-decision decoding with ARQ — 145
- 6.4.5 Chase's soft-decision algorithm — 146
- 6.4.6 Viterbi decoding of block codes — 147

Exercises — 148

7 Codes based on fields and polynomials — 151

7.1 Fields — 151
- 7.1.1 Modular arithmetic — 151
- 7.1.2 Rings — 152
- 7.1.3 Fields — 153
- 7.1.4 Extensions and $GF(2^m)$ — 155

7.2 Hamming and BCH codes — 157
- 7.2.1 Hamming codes — 157
- 7.2.2 BCH codes — 159

7.3 Polynomials — 160
- 7.3.1 Polynomials and their degrees — 160
- 7.3.2 Factors and multiples — 161
- 7.3.3 Division algorithm — 161

		7.3.4 The mod operation and congruences	162
		7.3.5 Polynomial rings	163
	7.4	Cyclic codes	163
		7.4.1 Generator polynomial	163
		7.4.2 Cyclic property	165
	7.5	Reed–Solomon codes	167
		7.5.1 The code	167
		7.5.2 Check-sum conditions	168
	7.6	Euclid's algorithm	169
		7.6.1 The algorithm	169
		7.6.2 Some consequences	171
		7.6.3 Properties of inverses mod G	173
		7.6.4 Examples of inverses	173
	7.7	Algebraic decoding	174
		7.7.1 Introductory	174
		7.7.2 The key equation	174
		7.7.3 Euclidean algorithm	176
		7.7.4 Erasures as well	176
		7.7.5 Variations on a theme	177
	7.8	Concatenated codes	178
		7.8.1 Double encoding	178
		7.8.2 Decoding complexity	180
		7.8.3 Reliability	180
	Exercises		182
8	**Optical communications and quantum effects**		**184**
	8.1	Optical communications	184
		8.1.1 Introductory	184
		8.1.2 Optical fibres	185
	8.2	Reception	187
		8.2.1 The photo-electric effect	187
		8.2.2 Photodetection followed by amplification	188
		8.2.3 The ideal photodetector	189
		8.2.4 PPM system	190
		8.2.5 Coherent reception	192
	8.3	Quantum theory	195
		8.3.1 Photons in an ideal single-mode cavity	195
		8.3.2 The suppression of thermal noise	196

	8.3.3	Maximum rate of transmission of information	197
	8.3.4	More on photon counting	199
	8.3.5	Comparisons with the upper bound	200
	Exercises		201

9 Cipher systems 203

9.1 Introduction 203

9.1.1	Basic terminology	203
9.1.2	Monoalphabetic and bigram ciphers	204
9.1.3	Running-key cipher	205
9.1.4	Some further comments	205
9.1.5	Functions and mappings	206

9.2 Conventional ciphers 207

9.2.1	The one-time pad	207
9.2.2	Theoretical secrecy	208
9.2.3	Known-plaintext attack	210

9.3 Stream ciphers 211

9.3.1	Introductory	211
9.3.2	Feedback shift-registers	211
9.3.3	The division algorithm revisited	212
9.3.4	Introducing non-linearity	214
9.3.5	Other problems	215

9.4 Block ciphers 216

9.4.1	Affine transformations	216
9.4.2	IBM systems	217
9.4.3	Feistel cipher and the DES system	219
9.4.4	Cipher feedback	220

9.5 Public key ciphers 221

9.5.1	Public key-distribution	221
9.5.2	RSA cipher	223

Exercises 226

Appendixes 228
A Exact error probability for biorthogonal codes 228
B Justification of the Euclidean algorithm 229

References 232

Index 236

Glossary

$A(f)$	two-sided noise spectral density, p. 79
$a(t)$	autocorrelation function of noise, p. 79
B	bandwidth equivalent, p. 95
C	channel capacity (Ch. 5), (5.42) on p. 113
C'	scaled channel capacity, $= C \ln 2$, (5.43) on p. 113
$\text{cov}(x)$	covariance of random variable x, (4.5) on p. 72
d	Hamming distance (Ch. 6), p. 133
E	energy of signal, p. 88
E_b	energy per bit of information, (6.14) on p. 135
$\text{erfc}(x)$	error-function complement, (4.18) on p. 76
$\text{erf}(x)$	error function, p. 228
f	frequency, p. 6
GHz	frequency unit, $= 10^9$ Hz, p. 7
h	Planck's constant, p. 187
$H(f)$	filter transfer function, p. 22
$h(t)$	filter impulse-response function, p. 21
h_{ij}	parity-check matrix (Chs. 6, 7), (6.18) on p. 137
I	information, (3.1) on p. 40
J	information rate, p. 88
J'	scaled information rate, $= J \ln 2$, (4.56) on p. 88
kHz	frequency unit, $= 10^3$ Hz, p. 6
k_B	Boltzmann's constant, p. 10
MHz	frequency unit, $= 10^6$ Hz, p. 6
mod	mod symbol, p. 123, p. 162
$\bar{n}$	photons per normal mode (Ch. 8), p. 200
P	signal power, p. 87
$P(A)$	probability of event A, p. 70
$P(A\|B)$	probability of event A given event B, p. 71
P_b	bit-error probability, p. 143
P_e	error probability, (4.52) on p. 87
P_W	word-error probability, p. 142
R	code-rate, $= k/n$, (6.12) on p. 135
$R(t)$	raised-cosine function, (2.6) on p. 15
$\text{sinc}(x)$	sinc function, $= \sin(\pi x)/(\pi x)$, (2.5) on p. 14

t	number of errors (Chs. 6, 7), p. 135
T	period, p. 26, or long interval of time, p. 45 n.
T_b	duration of pulse or bit, p. 45 n.
T_W	duration of codeword, p. 45 n.
THz	frequency unit, $= 10^{12}$ Hz, p. 7
var(x)	variance of random variable x, (4.4) on p. 72
W	bandwidth, p. 26
w	Hamming weight (Ch. 6), p. 134
z	polynomial argument (Ch. 7), p. 160
$z(t)$	positive-frequency representation, (3.17) on p. 61
α	field element obeying $\alpha^4 = \alpha + 1$ (Ch. 7 only), p. 156
α	subscript denoting member of ensemble (Ch. 4), p. 72
γ_b	$= E_b/\eta$, (6.13) on p. 135
$\delta(t)$	delta function, p. 19
δ_{kl}	Kronecker delta symbol, (5.6) on p. 96
η	(η_{NE} in Ch. 8) white-noise density, (4.36) on p. 80
η_Q	(Ch. 8 only) quantum efficiency, p. 189
θ_N	noise-temperature p. 82
$\phi(t)$	phasor, (3.21) on p. 62
$\Pi(t)$	rectangular-pulse function, (2.3) on p. 14
$\rho(x')$	probability density, p. 73
$\rho(x',y')$	joint probability density, p. 73
$\equiv$	congruence symbol, p. 162
$*$	convolution symbol, p. 18
(n, k) code	p. 123
$\binom{n}{s}$	combinatorial factor, p. 71
$\langle x \rangle, \bar{x}$	mean of random variable x, p. 72
$\text{III}(t)$	sha function, (2.54) on p. 35

1. Introduction

1.1 Aims and outline

1.1.1 *Introductory comments*

This book is planned around a theme, the theme of reliable digital communications. This theme is used as a guide for selecting topics from communications theory. It is not rigidly pursued however, and side-tracks of interest are often investigated as well. In consequence the text covers both the basics of digital communications and the basics of the theory of error-correcting codes as a single topic. It provides an introduction to coding theory in the context of traditional communications theory. This has been done by introducing coding as a form of digital modulation. In consequence it will be found that this book, although it starts in the manner of most textbooks on communications theory, brings in considerations of reliability and coding at a fairly early stage. The words 'communications', 'digital', and 'reliable' are examined further in the next three subsections.

The second part of the title refers to error-correcting codes. These are becoming more widely used in communications for at least the following reasons:

(a) The electronic hardware for carrying out the necessary computations for encoding and decoding is becoming cheaper and more reliable, so that what a few years ago was not economically feasible is easily carried out now.

(b) Some applications, such as the sending of data by *cipher*, need a level of reliability higher than usual. Ciphers are designed to detect tampering with the data-stream and so even a slight error may well leave a block of data indecipherable.

(c) The alternative of using error-checking, followed if necessary by a request for retransmission, may be clumsy where the round-trip time is rather long, as on a satellite link, or it may be inconvenient or even impossible, as with a recording or a broadcast sent to several receivers.

(d) Modern developments such as fibre-optic cables provide channels which can bear the extra load imposed by error-correcting codes. These codes always increase the amount of data to be sent since they work by introducing redundant checks by means of which errors are not only detected but also corrected.

1.1.2 *Communications*

The word 'communications' will refer to the point-to-point transmission of information, usually by electrical means, either by wire, coaxial cable, waveguide, or by radio. The overall system consists of several parts:

(a) The first component is the source, which provides the information to be sent. There are basically two types of information-signal, analogue and digital. Analogue signals are in the form of a continuously varying function of time. Typical sources of such signals are microphones and television cameras, which translate sound- and light-intensities respectively into electrical signals. Digital sources may be computer files or impulse streams used for driving remote teletypes or visual display units. Here the information is *coded* into a stream of pulses, whose heights can take only a discrete set of values, usually just two.

(b) The output from the source is then processed by a modulator in preparation for sending it over the channel.

(c) The channel is the link to the remote location where the information is to be sent. It may be a wired electrical connection, such as a telephone or telex line, or a radio link, where the transmitter radiates electromagnetic waves into space which are picked up by a receiver at the remote location. Other examples of channels are information-storage systems such as magnetic discs and tapes, which can be regarded as linking the sender who puts in a message at one time to the receiver who retrieves it, perhaps months later. The channel is the part of the communication system not under the control of the sender or receiver, and it tends to degrade the signal. Two main forms of such degradation will be considered. The channel can distort the signal in a predetermined way, which can be compensated for. It can also introduce interference, or *noise*, into the received signal in a manner which is unpredictable except in a statistical sense.

(d) The channel delivers the signal to the receiver. The degradation can cause errors at the receiver, and quite sophisticated means may be needed to combat its effects.

(e) After any necessary processing the signal is passed on from the receiver to its destination, the *information sink*. What happens to it after that is not the concern of this book.

1.1.3 *The robustness of digital communications*

Digital signalling has an innate capacity for countering interference which gives it a quality of robustness or strength. We illustrate this by contrasting digital and analogue signalling.

Suppose that a remotely situated instrument has to send back the values of some physical variable such as wind-speed or temperature to a central station along a signal-wire. A very simple method is to put a voltage on the wire proportional to the instrument reading. The voltage is then an analogue of the physical quantity being measured and so this system of sending the results is called *analogue* transmission. The possible values of the physical quantity and hence the corresponding voltage form a continuous range. Hence if there is any interference on the wire which changes the voltage as measured at the receiving end there is no simple way of knowing that the received value is false.

Now suppose the transmitter is designed to send voltages from a discrete set of values, say, 0, 1, ..., 9. If the interference on the wire is very unlikely to cause the output to deviate by more than 0.5 V then the receiver can tell quite reliably what actually was sent. Thus if 7.21 V is received then it is almost certain that 7 V was sent. Only very occasionally when the interference happens to be too large is the receiver fooled. A transmission system, in which the possible signals fed into the channel form a discrete set, is said to use *digital signalling*. As noted above, this discreteness provides a capacity for correcting the effects of interference and hence can be used to make the signalling very reliable. This advantage of digital signalling over analogue signalling may not be very great in a simple system, but if the signal has to go through a lot of complicated processing then the tendency of analogue signals to 'drift' makes them difficult to deal with. In contrast digital signals can be corrected or re-formed at each stage.

There are two main disadvantages of digital signalling. The first is that the electronics may be rather complicated. This is no longer necessarily a serious disadvantage, since developments in electronics have made available hardware of incredible complexity combined with remarkable cheapness and reliability against breakdown. The second disadvantage is that digital signalling makes greater demands on the channel. (In particular it needs a greater bandwidth.) Thus suppose that we wish the remote instrument to send its readings accurate to three figures. Sending a reading like 7.52 is no harder than sending a single integer for an analogue system, but the digital system just described has to send three distinct digits, 7, 5, and 2, either in succession (serially) along the wire, or in parallel along three wires. Again, developments in the techniques of communication have reduced the seriousness of this problem.

The points made in this section can be illustrated by considering the trend towards digital sound-recording (Philips Technical Review 1982). In the old system the sound-signal was represented in an analogue manner by the continuously variable magnetization of a magnetic tape or by the lateral displacement of the groove on a record. Digital recording involves sampling the signal very frequently (at 44.1 thousand times a second), and representing the sampled values digitally to a sufficiently high precision. (Sixteen binary digits are used rather than the three decimal digits in the example above.) The ensuing stream of digits is then written on to a medium such as a video disc or video tape, originally developed for analogue television signals, and which in consequence can provide the extra resources (in bandwidth) needed. This technique can obviate not only the hiss and crackle caused by imperfections in the record, but also the effects of unsteadiness in the turntable speed, and frequency and intermodulation distortion in the recording process.

1.1.4 *Reliable communications*

If the reliability of a signalling system is too low, then one way of getting a message through is to send it several times, so that the receiver can piece it together from the various garbled versions received. Such a strategy evidently slows up the transmission rate, the rate at which messages are getting through, and it is easy to see that the achievement of very high reliability involves a great number of repetitions. It seems reasonable to expect that, in general, arbitrarily high reliability can only be achieved at the cost of an excessively low transmission rate. However, Shannon (1948) proved that this is not so, and that arbitrarily high reliability can be achieved while keeping the transmission rate up at a finite value. His work spurred on the development of both information theory and of coding theory. Information theory provides precise definitions of the amount of information and of information rates. Shannon proved that arbitrarily high reliability can be achieved if and only if the rate of transmission of information is below a rate called the *channel capacity*. This reliability is achieved by enhancing the robust quality of digital signalling just described. In practice error-correcting codes are used and the attainment of very high reliabilities is part of the province of coding theory. Unfortunately, Shannon's proof failed to demonstrate a practical coding scheme, and in consequence it was believed for a while that any code achieving the levels of reliability promised by Shannon would be impossibly complicated and totally impractical to use, employing huge lookup tables whose entries might well exceed the total number of atoms in the earth. However, as we shall see, families of codes have been found which can achieve arbitrarily high reliability without excessive complexity. In consequence, some of these codes are useful in practice; they are not too hard to decode, and yet they can attain reliabilities as high as one may reasonably demand in this uncertain world.

1.1.5 *Outline*

The rest of this chapter will be used to fill in background material.

Chapter 2 contains a brief summary of Fourier theory, one of the mathematical tools needed in communications theory. One may remind oneself of how basic it is by noting that the concepts of frequency and of bandwidth in radio signalling are derived from Fourier theory.

Chapter 3 discusses ways in which digital signalling is carried out. It draws heavily on the Fourier theory of Chapter 2 in the discussion of bandwidths of pulsed signalling. Two examples of the robustness of digital signalling are given. In the first example the probability of error is considered when a signal has to be relayed by a long line of repeaters, and in the second it is shown how adaptive equalizers can be made to track frequency and phase errors in the transmission. Another important topic is how analogue signals can be faithfully sampled for digital transmission or recording. It seems reasonable to expect that sampling would cause significant degradation, but in fact this need not be so. The topics of information theory and data-compression are touched on, but only superficially since that is sufficient for the discussions at a later stage.

In Chapter 4 a second mathematical tool is introduced, probability theory. This is used to build up a simple and useful model of a noisy channel, the Additive White Gaussian Noise (AWGN) channel. A simple method of pulsed signalling is suggested which can achieve arbitrarily high levels of reliability at finite rates of transmission of information, but unfortunately its use involves excessive peak power at the transmitter, excessive bandwidth on the channel, and excessive complexity at the receiver. These difficulties are circumvented one by one in the course of the next three chapters.

In Chapter 5 the idea is further developed that the discreteness of digital signals provides their robustness. The different pulse-levels allowed may be plotted as points on a linear scale, and so form a discrete set of points along a line, a one-dimensional space. This geometrical concept is then generalized to show how the codewords of an error-correcting code may be regarded as points in a multidimensional space, whose spacing provides the code with its robustness and error-correcting abilities. Certain types of code are considered in this manner. The discussion leads naturally on to Shannon's theorem (specialized for the AWGN channel), and its derivation.

This approach to coding theory is a little unconventional. The usual introduction is in terms of extra check symbols, which are introduced into the data by an *encoder* at the transmitting end. At the receiving end a *decoder* has to decide from the pattern of check-failures what was actually sent. In Chapter 6 error-correcting codes are discussed from this point of view. Their performance on the AWGN channel is considered. Other types of channel such as the *fading* channel are mentioned, although there is not enough space to deal with them in much detail. Shannon's theorem is used to find out what is ideally available from coding, and whether it is worth using coding at all to increase the reliability of digital communications.

In Chapter 7 the development of coding theory is continued by the introduction of a third mathematical tool, that of the algebraic theory of fields. In this way some new types of code are introduced, but more importantly the theory gives a very powerful error-correcting technique suitable for such codes. A combination of the ideas in this chapter with those in the previous two chapters provides families of codes which can attain the reliability promised by Shannon, without the problem of excessive compexity in the error-correcting process at the receiver.

The increasing importance of optic-fibre communications points towards a study of the implications of quantum theory for reliable communication. At optical frequencies various quantum effects begin to show up which are drowned at lower frequencies by various forms of noise such as thermal noise. The basics of communication by optic fibres are discussed in Chapter 8, as well as the fundamental limits on information rates imposed by quantum theory.

Finally in the last chapter a different aspect of reliable signalling is discussed, that of obtaining security against eavesdroppers. This leads into the topic of cryptology, the study of ciphers. Here some ideas developed in Chapter 3 on information theory are further made use of. Important topics introduced here are the subject of public-key ciphers and the idea that algorithms for encryption may be safely published.

1.2 Physical background

1.2.1 *Frequency bands*

An oscillation of the simple harmonic form $\cos(2\pi ft)$ is said to have a *frequency* f Hz, and its graph goes through f oscillations or cycles in 1 second. The duration of each cycle f^{-1} is called the *period*. The *wavelength* λ is the distance the wave-train causing the oscillation travels in one period. If the speed is c then

$$\lambda = c/f. \tag{1.1}$$

Electromagnetic waves travel in vacuum at the speed of light $c = 2.9979 \times 10^8$ m s^{-1}. and at almost this speed in the earth's atmosphere. (Tables of physical constants are given in Kaye and Laby 1972, and in the *American Institute of Physics Handbook* 1972.) Thus an electromagnetic wave with a wavelength of 100 m has a frequency of 2.9979 MHz (1 MHz = 10^6 Hz), usually approximated as 3 MHz.

A radio station is assigned a carrier frequency f_c. If there is no information to be sent it transmits a signal proportional to $\cos(2\pi f_c t)$; this *carrier* is varied or *modulated* in some way when information is transmitted. A typical method is to send an *amplitude modulated* (AM) signal $A(t)\cos(2\pi f_c t)$, where the amplitude $A(t)$ is varied to convey the information to be sent. The effect of this modulation is to spread the frequency of the transmitted signal into a band centred on f_c, the width of this band being of the order of the reciprocal of the time-scale in the variation of $A(t)$. The required *bandwidth* is of the order of 10 kHz (1 kHz = 1000 Hz) for sound broadcasting, and 10 MHz for television. The bandwidth of channels conveying information is proportional to the rate at which information is sent. The carrier-frequency should exceed this bandwidth. It usually does so, by a large factor, in which case the transmission is called *narrow-band*. Thus commercial AM sound broadcasting is mostly carried out at frequencies well in excess of 100 kHz, and television is sent with carrier-frequencies above 40 MHz. A receiver picking up a signal with a carrier frequency f_c contains *bandpass filters* which accept signals in the appropriate frequency band around f_c and which reject signals at all the other frequencies. (For details of how frequencies are assigned in the USA, see ITT 1975.)

The range of frequencies used for electromagnetic communications extends over many orders of magnitude, from about 10^4 Hz to 10^{15} Hz. (The corresponding wavelengths are respectively 30 km and 300 nm.) (1 km = 10^3 m, 1 nm = 10^{-9} m.)

Some of the reasons for going to higher frequencies are:

(a) the low-frequency channels are already allocated,
(b) large bandwidths are needed for high-speed information transmission, so that the carrier-frequencies need to be high, and
(c) antennae can be made highly directional so that there is more power at the receiver and less available to cause interference with other users of the same frequency, and to be picked up by an eavesdropper.

Radio waves at frequencies below about 30 MHz are reflected down by the ionosphere, a series of ionized layers in the upper atmosphere, and in consequence can travel thousands of kilometres by bouncing between the earth and the ionosphere. The ionosphere is transparent to waves at frequencies above about 30 MHz, so that long-distance terrestrial communications are not normally possible in this range. However, these frequencies are instead used for communicating with satellites above the ionosphere. Terrestrial radio communication has an upper limit of about 100 GHz (1 GHz = 10^9 Hz) because of atmospheric absorption and scattering. Satellite-to-satellite communications can in principle take place at frequencies up to several hundred THz (1 THz = 10^{12} Hz).

The range of possible frequencies from 3 kHz to 300 GHz is traditionally broken up into bands (Martin 1978, Chapter 2). These are listed below, together with some of their properties of interest for terrestrial communications:

(a) 3 to 30 kHz. Very-low-frequency (VLF) band. Electromagnetic waves at these frequencies are used to communicate with submarines, since they penetrate a few metres into sea-water (Callahan 1981). Transmitting antennae are indeed huge structures, since they should have dimensions of at least 1/10 of the wavelength which exceeds 10 km.

(b) 30 kHz to 300 kHz. Low-frequency (LF) band. Used by some 'long wave' European broadcasting stations.

(c) 300 kHz to 3 MHz. Medium-frequency (MF) band. Used for 'medium wave' broadcasting. Waves in this band are absorbed by the ionosphere rather than reflected, especially in the daylight hours, and so the range is limited.

(d) 3 MHz to 30 MHz. High-frequency (HF) band. Used for long-distance 'short wave' communications.

(e) 30 MHz to 300 MHz. Very-high-frequency (VHF) band. Used for mobile communications, television, and wide-band frequency-modulated (FM) sound broadcasting.

(f) 300 MHz to 3 GHz. Ultra-high-frequency (UHF) band. Used for mobile communications and television. Communications can take place only along a line of sight and strong shadows are thrown by hills and buildings.

(g) 3 GHz to 30 GHz. Super-high-frequency (SHF) band. Used for wide-band point-to-point communications. (Note that the term 'wide-band' just means a large bandwidth and does not necessarily say anything about the bandwidth divided by the carrier-frequency.) Frequencies around 12 GHz have been allocated for direct broadcasting by satellite direct into people's homes (DBS).

(h) 30 GHz to 300 GHz. Extra-high-frequency (EHF) band. Used for short-range communications, which are easily disturbed by rain.

The term 'microwave' loosely covers frequencies from about 1.5 GHz to 30 GHz, and 'millimetre waves' from 30 to 300 GHz. Above 300 GHz it is possible to designate bands roughly as follows, although there are no strict definitions:

— 300 GHz to 3 THz, submillimetre waves
— 3 THz to 30 THz, far-infrared waves
— 30 THz to 300 THz, near-infrared waves
— 300 THz to 1000 THz, visible radiation and its vicinity.

Frequencies around 200 to 350 THz are used for fibre-optic communications. The quantum nature of radiation becomes apparent at such frequencies. Electromagnetic radiation from 430 THz (red light) to 750 THz (violet light) is directly visible and is used for all sorts of human communications, but not usually for wide-band electronic communications.

1.2.2 *Power and power flux*

If a source radiates power P, then the power flowing through the surface of a sphere of radius R centred on the source is distributed over the surface area $4\pi R^2$ of the sphere. If the source is *isotropic* so that the power is distributed uniformly, then the power flux Φ, the flow of power through unit area perpendicular to the flow, is given by

$$\Phi = P/(4\pi R^2). \tag{1.2}$$

It is measured in watts per square metre (W m^{-2}). In terms of the electric field E associated with the wave the power flux is $\Phi = E^2/Z_0$, where $Z_0 = 376.7$ ohms is the 'impedance of free space'. Thus a 1 kW transmitter (typical for a local broadcasting station) produces a power flux of 7.958×10^{-7} W m^{-2} at 10 km, and an electric field of 17.31 mV m^{-1}.

A transmitting antenna may be directional, so that more power is sent in some directions than in others. Then the power flux at the receiver can be increased by a factor called the *gain* of the antenna. For a dish antenna consisting of a parabolic reflector behind the antenna proper (such as is used at microwave frequencies) the gain G is given by

$$G = 4\pi A_T/\lambda^2 \tag{1.3}$$

where A_T is the effective area of the dish, the product of its geometric area and an efficiency factor (of about 0.55) (Martin 1978, Chapter 6). Here λ denotes the wavelength. A similar dish antenna used at the receiver when correctly aligned delivers a power

PHYSICAL BACKGROUND

$$P_R = \Phi A_R \tag{1.4}$$

to the receiver, where A_R is the effective area of the receiving dish. Again this effective area is the product of the physical area and of an efficiency factor (of about 0.55). Overall the received power is given by

$$P_R = P_T \cdot A_R A_T/(\lambda^2 R^2). \tag{1.5}$$

Of course this applies only if the antennae are pointed correctly at each other.

The product GP_T is called the effective radiated power (ERP) of the transmitter. Broadcast transmitters at VHF and UHF frequencies direct their power horizontally, to stop power being sent uselessly into the ground and into outer space. Large television transmitters have effective radiated powers of the order of 100 kW in the VHF band and 1000 kW in the UHF.

Electrical signals at the receiver are commonly measured in volts, with the power flowing into the receiver given by the square of the voltage divided by the 'input impedance' of the receiver. Since this impedance varies from receiver to receiver, the square-root of a watt will be chosen as the unit for signal-amplitude, so that the square of the amplitude gives the power directly.

1.2.3 *The decibel*

The designer of a system often has to work out expressions like (1.5) for power levels, which involve a large number of multiplications and divisions of quantities with outlandish magnitudes. It seems natural to use logarithms and the convention is to take logarithms to the base 10, and then multiply by 10. The unit of power when processed thus is called the *decibel* (dB).

(a) A power P_2 is said to be d dB above (or $-d$ dB below) a power P_1 with

$$d = 10 \log_{10}(P_2/P_1). \tag{1.6}$$

Thus if the power input to an amplifier (or attenuator) is P_1 and the output is P_2, then the gain is said to be d dB, and if this is negative it may be said that the attenuation is $-d$ dB.

(b) Often P_1 in (1.6) is given a reference value of 1 W. Then P_2 is said to be d dBW (decibels relative to a watt). Similarly for dBm (decibels relative to a milliwatt) and dBf (decibels relative to a femtowatt = 10^{-15} W). For example a power of 2 μW is -57 dBW, or -27 dBm, or 93 dBf.

(c) Power fluxes are also measured in dB. In particular sound intensities (which are power fluxes) are given in dB relative to 10^{-12} W m^{-2}, which is taken as the 'threshold of hearing'. Thus a sound level of 65 dB (typical of normal conversation) is just over 3 μW m^{-2} (ITT 1975, Chapter 37).

(d) The apparent luminosity of a star is also measured as a power flux (Allen 1973). A star of magnitude 0 gives a light-power flux of about 3.73×10^{-9} W m^{-2}. A star of magnitude n gives a flux $-4n$ dB relative to this value, so that a sixth-magnitude star (which is just visible without instruments) is −24 dB relative to the flux just quoted.

1.2.4 *Noise and interference*

Electromagnetic waves produced naturally tend to occupy a large bandwidth. Of course they interfere with genuine signals at the receiver and so they are a form of *interference* or *noise*. One particular source of chaotic electromagnetic waves is a hot body, and if it is hot enough (red hot or white hot) this radiation becomes visibly apparent. Such radiation is called *thermal* radiation. The amount of noise-power N picked up by a narrow-band receiver is proportional to the bandwidth W of the receiver, that is the width of the band of frequencies the receiver accepts. The strength of the noise is then measured as the ratio $\eta = N/W$ and has the units of watts per hertz or simply of joules. In particular the radiation inside a box with opaque walls at absolute temperature θ has a value of η given by $\eta = k_B \theta$, where $k_B = 1.38 \times 10^{-23}$ J/K (joules per kelvin) is Boltzmann's constant. This radiation is known as *black-body radiation*. (Absolute temperatures in degrees kelvin (K) are given by adding 273.2 to centigrade temperatures.) Thus at a temperture of 290 K (27 C) the value of η is 4×10^{-21} J. In that case the noise-power picked up by a receiver with a bandwidth of 10 MHz is 4×10^{-14} W.

Other forms of random behaviour at the front end of a receiver produce additional noise whose power is proportional to the bandwidth. Such noise is also measured as an energy, or as a *noise-temperature*, this energy divided by k_B. Typical values for noise-temperatures are around 1000 K, but can be as low as 0.5 K for specially designed receivers operating at low temperatures. This kind of noise in a domestic black and white television receiver produces hiss on the sound-channel and a 'snowstorm' on the picture.

The noise-energy η or noise-temperature η/k_B for true thermal noise is independent of frequency up to very high frequencies (see Chapter 8), but may vary considerably with frequency for other sources of noise. Thus the natural background for radio communications has a value of the order of 10^4 K (when expressed as a temperature) in the LF band, but drops to around 10 K in the region of 1 to 10 GHz (ITT 1975). Above these frequencies it starts to climb again towards 300 K. The natural noise below 20 MHz is caused mainly by atmospheric effects such as lightning, and from 20 MHz to 10 GHz it is mainly caused by radiation from outer space, which provides the radio astronomer with his trade but is a nuisance in communications work. The noise above 10 GHz is mainly produced because the atmosphere acts like the walls of a partially opaque box surrounding the receiver. The fact that the atmosphere absorbs radiation at these frequencies means that the radiation is coupled to molecular vibrations, and so loses its energy to the molecules. But this also means that the random thermal motion of the molecules is electromagnetically active and so radiates interference.

PHYSICAL BACKGROUND   11

The designer of a communications system has to ensure that the signal-power S at the receiver exceeds the noise-power N by a sufficiently large factor. This factor, the signal-to-noise ratio, is usually expressed in dB. Thus for a decent picture on a television receiver this ratio should be 10^5 or more (i.e. $\geqslant 50$ dB). If the noise temperature θ_N is 1000 K and the bandwidth 10 MHz, then the noise power $N = Wk_B\theta_N$ is equal to 1.38×10^{-13} W, so that the signal must exceed 1.38×10^{-8} W.

So far only natural sources of interference have been described, and these are in most cases statistically fairly predictable. However, as any short-wave listener knows, most interference below 50 MHz is man-made. Such 'other-user' interference is not usually very predictable statistically, and nothing much will be said about it here. It is of course not purely a technical problem. Such interference can be a serious problem at higher frequencies as well.

For further reading: The books by Martin (1976, 1977, and 1978) provide much background material for the fascinating subject of electronic communications.

Exercises

1.1. Estimate the number of atoms in the earth.

1.2. A geostationary satellite is a satellite in orbit over the equator and which rotates with the earth once every 24 hours, so that it apparently always stays in the same place in the sky. Calculate the height of such a satellite above the earth's surface and the time it takes a signal to travel from the nearest point on the earth to the satellite and back. (Acceleration due to gravity at distance r from the centre of the earth $= gR^2/r^2$ where $g = 9.81$ m s^{-2} and where $R = 6.37 \times 10^6$ m is the radius of the earth. This must be equal to the centrifugal acceleration $\omega^2 r$ where ω is the angular velocity of the satellite in radians per second, corresponding to one revolution every 24 hours.)

1.3. Estimate the time-scale of the fastest changes in the level of a television signal, if a picture is scanned 25 times each second with 625 lines, and details in the picture of the size of 1/800 of a line-length can be resolved.

1.4. In the European television system the whole picture is built up from 625 lines, once every 1/25th of a second. The television signal travels from the transmitter to the receiver along two separate paths of different lengths. The delayed signal produces a 'ghost' on the screen 1/20th of a screen-width to the right of the main object. Estimate the difference in the lengths of the two paths.

1.5. Calculate the power picked up from a 20 W transmitter at a distance of 10^{12} m, operating at a frequency of 10 GHz. Assume that the transmitting antenna is a dish of effective area 6 m^2 and the receiving antenna a dish of effective area 2000 m². (These values are typical for the communications link of a mission to the outer planets. See Edelson, Madson, Davis, and Garrison 1978.)

1.6. What is the efficiency of a loudspeaker which gives a sound level of 96 dB at 1 m with an electrical input of 1 W? Assume that the loudspeaker radiates isotropically.

1.7. Calculate the power radiated by an isotropic source of sound which produces a sound-level of 65 dB at 1 m. (This is typical of the human voice in normal conversation.)

1.8. Calculate the power needed by a satellite to broadcast a television signal over an area of 10^6 square kilometres, assuming that the effective area of a receiving dish is 1 m², that the bandwidth is 10 MHz, and that the signal level should be 50 dB above the noise which corresponds to a temperature of 300 K.

1.9. The sun has a magnitude of -26.5. Calculate the power-flux of sunlight in W m^{-2}. Do the same for the full moon, with a magnitude of -12.5.

2. Fourier theory

The basic ideas behind Fourier theory are summarized in this chapter, in anticipation of their use later on. There is no serious discussion of questions of rigour, concerning which the reader is referred to the monographs of Lighthill (1959) and Bracewell (1978). The latter also provides a more extended introduction to the subject.

Fourier integrals and the Fourier inversion theorem are considered first of all in Section 2.1. This order of presentation, with Fourier integrals being discussed before Fourier series, follows Lighthill and Bracewell. The main objection to this arrangement is the lack of any obvious motivation behind the Fourier inversion theorem. On the other hand this order seems logically reasonable and is the order in which the parts of the theory are used in this book. The idea of signals occupying frequency bands, used a bit vaguely in Chapter 1, is made more definite. Various examples of Fourier transforms are discussed, as well as their general properties, and the discussion leads on to the convolution theorem, which is vital in understanding the operation of filters. The delta function (impulse function) is introduced in Section 2.2, and used in developing the theory of filters a little further.

Fourier series for periodic functions are introduced in Section 2.3. Periodic functions are highly idealized since they are supposed to go on for ever, but are none the less a very convenient concept.

Finally periodic delta-comb functions are introduced in Section 2.4 as a way of deriving the Nyquist theorems. These are the basis for the theory of the digital processing of analogue signals.

2.1 The Fourier transform

2.1.1 *The Fourier inversion theorem*

The function $V(f)$ defined by

$$V(f) = \int_{-\infty}^{\infty} v(t) e^{-2\pi jft} dt \qquad (2.1)$$

is called the *Fourier transform* of $v(t)$. Here $v(t)$ is a function of the time t usually representing a signal, the real variable f is the frequency, and j is the square-root of -1. In general V is a complex quantity, and there is no harm in allowing $v(t)$ to be complex as well. We suppose that $v(t)$ is such that the integral converges for every value of f from $-\infty$ to $+\infty$. As far as possible in the general theory lower-case letters will be used for functions of time, and the corresponding capitals for their Fourier transforms. The Fourier transform is important since it enables us to represent $v(t)$ as

$$v(t) = \int_{-\infty}^{\infty} V(f) e^{2\pi jft} df, \qquad (2.2)$$

provided $v(t)$ obeys certain conditions to be discussed later in this subsection. The Fourier inversion theorem states that the right-hand side of (2.2) is indeed equal to $v(t)$ for all real values of t. We call $v(t)$ the *inverse Fourier transform* of $V(f)$.

We may regard (2.2) as showing how the signal $v(t)$ is built up from 'frequency components' $V(f)e^{2\pi jft}$. For this reason $V(f)$ is often referred to as the *frequency spectrum* of the signal $v(t)$. In particular if $V(f)$ is negligible for $|f|$ greater than some specified value f_M we call $v(t)$ band-limited and say that it contains no frequencies greater than f_M. Similarly if $V(f)$ is negligible unless $|f|$ lies in a specified band of frequencies, then $v(t)$ is said to be a signal *occupying* that band of frequencies. It is mathematically more convenient to use the complex exponential rather than the cosine and sine functions which are respectively its real and imaginary parts. The cost of doing this is that the integral has to be taken over negative values of f (negative frequencies) as well.

We now consider a few examples of Fourier transforms, listed in Table 2.1. The first entry needs a little explanation. The function $\Pi(t)$ is defined by

$$\Pi(t) = \begin{cases} 0 & \text{if } |t| \geqslant \tfrac{1}{2} \\ 1 & \text{if } |t| < \tfrac{1}{2}. \end{cases} \qquad (2.3)$$

It is a rectangular pulse of unit width and unit height. Its Fourier transform is

$$V(f) = \int_{-\frac{1}{2}}^{\frac{1}{2}} e^{-2\pi jft} dt = \sin(\pi f)/(\pi f). \qquad (2.4)$$

This function is so common that we give it a special name and define

$$\operatorname{sinc}(x) = \sin(\pi x)/(\pi x). \qquad (2.5)$$

(Note the factor π.) This function is unity for $x = 0$ and vanishes for every other integral value of x.

TABLE 2.1

$v(t)$	$V(f)$
$\Pi(t)$	$\operatorname{sinc}(f)$
e^{-at} if $t > 0$, 0 if $t \leqslant 0$ $(a \geqslant 0)$	$1/(a + 2\pi jf)$
$e^{-a|t|}$ $(a \geqslant 0)$	$2a/(a^2 + 4\pi^2 f^2)$
$1 - |t|$ if $|t| \leqslant 1$, 0 if $|t| > 1$	$\{\operatorname{sinc}(f)\}^2$
$R(t) = \{1 + \cos(\pi t)\}/2$ if $|t| < 1$, 0 if $|t| \geqslant 1$	$\sin(2\pi f)/[(2\pi f)\{1 - (2\pi f)^2\}]$
$\exp(-\pi t^2)$	$\exp(-\pi f^2)$

Other Fourier transforms are tabulated in Table 2.1, including the Fourier transform of the *raised cosine* function

$$R(x) = \begin{cases} \frac{1}{2}\{1 + \cos(\pi x)\} & \text{if } |x| < 1, \\ 0 & \text{if } x \geq 1, \end{cases} \quad (2.6)$$

which will be used in Chapter 3. The functions in this table become progressively smoother as we go down the list. Thus the first two examples of $v(t)$ contain discontinuities, the third and fourth are continuous functions with discontinuities in the first derivative, the fourth and fifth functions are continuous with continuous first derivatives but have discontinuities in the second derivative. The sixth function has all its derivatives continuous in the whole range from $t = -\infty$ to $t = +\infty$. Such a function will be called 'smooth'. In fact many elementary functions such as polynomials, $\sin(t)$, $\cos(t)$, and $\exp(t)$ are smooth in this sense. The Fourier transform of the sixth entry in Table 2.1 is a special case of the result

$$\int_{-\infty}^{\infty} \exp\{-a(t+b)^2 - c\} \, dt = \sqrt{(\pi/a)} \, e^{-c} \quad (2.7)$$

where a, b, and c are complex constants with $\text{Re}(a) > 0$. This follows from

$$\int_{-\infty}^{\infty} \exp(-x^2) \, dx = \sqrt{\pi}$$

(Exercise 2.2) by an appropriate substitution (and a shift in the contour of integration if a and b have non-zero imaginary parts; see Spiegel 1964).

We now consider the conditions for the validity of the Fourier inversion theorem. It certainly works for a class of functions called 'good' by Lighthill (1959). A 'good' function is an infinitely differentiable or smooth function with the property that it and all its derivatives tend to zero as t tends to infinity faster than any negative power of t. Thus $\exp(-\pi t^2)$ is a good function. So is the function

$$f(t) = \begin{cases} \exp\{-2/(1-t^2)\} & \text{for } |t| < 1, \\ 0 & \text{for } |t| \geq 1, \end{cases}$$

which is an example of a good function which is non-zero only on a finite interval (Exercise 2.5).

We would expect physically realizable signals to be good functions, being presumably both smooth and of finite duration. Functions like $\Pi(t)$ can be regarded as limiting cases of good functions, the limit never being reached in physical realizations of pulse-functions because rise-times must always be finite. By appropriately taking the limits it is possible to extend the Fourier inversion theorem to a class of functions called 'generalized functions'. This class includes all the functions in the above table, and so the theorem also applies to them as well. (To demonstrate (2.2) with some of the entries in this table is not straightforward and may require the techniques of contour integration.) In the class of generalized functions there are also functions which although not physically realizable are indispensable in the theory. These include (a) periodic functions (which cannot be physically realized since they go on for ever), (b) δ or impulse functions, which instantaneously become infinite in value, (c) periodic stream of δ-functions, also known as the III function or periodic delta comb, (d) derivatives of δ-functions. The impulse functions and their derivatives are not functions in the ordinary sense at all, but fortunately can be manipulated almost as readily as ordinary functions. On the other hand there are some non-periodic signals such as noise which also do not die away at infinity. These cannot be rigorously dealt with without using advanced techniques. Instead we shall resort to the device of forcing such functions to be periodic over a period much greater than any other time-scale in the discussion.

2.1.2 A few simple results

(a) As a simple but useful consequence of (2.1) and (2.2) we have

$$V(0) = \int_{-\infty}^{\infty} v(t)\,dt \tag{2.8a}$$

and

$$v(0) = \int_{-\infty}^{\infty} V(f)\,df. \tag{2.8b}$$

Thus $v(0)$ is the area under the curve of $V(f)$ and $V(0)$ the area under the curve of $v(t)$.

(b) If the Fourier transform of $u(t)$ is $U(f)$, then the Fourier transform of du/dt is $2\pi j f U(f)$, the Fourier transform of $u(t - t_0)$ is $\exp(-2\pi j t_0)U(f)$, and the Fourier transform of $u(at)$ is $|a|^{-1}U(f/a)$. These results and others are tabulated in Table 2.2.

(c) The reality condition is:

$$\text{If } v(t) \text{ is real,} \quad \text{then } V^*(f) = V(-f), \tag{2.9}$$

where the asterisk denotes complex conjugation. Most signals are represented by physical quantities such as voltages and hence must be real. This condition guarantees that the imaginary parts cancel out in an integral like the one on the right-hand side of (2.2).

(d) The evident symmetry between t and f in (2.1) and (2.2) leads to the following result: the Fourier transform of $U(t)$ is $u(-f)$. This remarkable symmetry between f and t often enables us to obtain new results from old by interchanging f and t, with due regard for the minus sign in the exponent of (2.1). Results related in this way are called 'duals' of each other.

THE FOURIER TRANSFORM

TABLE 2.2 *Fourier transforms*

$u(t)$	$U(f)$
$v(t)$	$V(f)$
$au(t) + bv(t)$	$aU(f) + bV(t)$
$u^*(t)$	$U^*(-f)$
$du(t)/dt$	$2\pi j f U(f)$
$U(t)$	$u(-f)$
$\exp(2\pi j f_0 t) u(t)$	$U(f - f_0)$
$2\cos(2\pi f_0 t + \theta) u(t)$	$e^{j\theta} U(f - f_0) + e^{-j\theta} U(f + f_0)$
$u(t - t_0)$	$\exp(-2\pi j f t_0) U(f)$
$u(at)$ $(a \neq 0)$	$\|a\|^{-1} U(f/a)$
$u(t)v(t)$	$\int U(f') V(f - f') df'$ or $U * V$
$\int u(t') v(t - t') dt'$ or $u * v$	$U(f) V(f)$
$\tau^{-1} \Pi(t/\tau)$	$\text{sinc}(f\tau)$
$\exp(-\pi t^2)$	$\exp(-\pi f^2)$
$\tau^{-1} \exp(-\pi t^2/\tau^2)$	$\exp(-\pi \tau^2 f^2)$
$\delta(t)$	1
$\delta(at) = \|a\|^{-1} \delta(t)$	$\|a\|^{-1}$
$\delta(t - t_0)$	$\exp(-2\pi j f t_0)$
$\exp(2\pi j f_0 t)$	$\delta(f - f_0)$
$\text{III}(t) = \Sigma_k \delta(t - k)$	$\text{III}(f)$
$c(t) = \Sigma_k \delta(t - kT) = T^{-1}\text{III}(t/T)$	$C(f) = T^{-1} \Sigma_n \delta(f - n/T) = \text{III}(fT)$

(e) Let $v(t)$ be the indefinite integral of $u(t)$:

$$v(t) = \int_{-\infty}^{t} u(t') \, dt'. \tag{2.10a}$$

Then $v'(t) = u(t)$ and hence the Fourier transforms should satisfy

$$V(f) = (2\pi j f)^{-1} U(f). \tag{2.10b}$$

It is safe to use this result only if the limit on the right-hand side as $f \to 0$ is finite and well-behaved. In particular by (2.8a) the area under the curve of $v(t)$ must be finite.

2.1.3 Convolutions and correlations

It may be shown that in manipulating generalized functions the orders of integration may be interchanged, and also that generalized functions possess a 'uniqueness property': if two generalized functions possess the same Fourier transform or inverse Fourier transform then the functions are identical. (In particular this implies that if the Fourier transform $U(f)$ (2.1) of some generalized function $u(t)$ is zero for all f, then $u(t)$ vanishes for all t. There are so-called null-functions which always give zero when integrated and yet are not themselves zero everywhere. However in the theory of generalized functions such functions cannot be distinguished from zero.)

The following theorem is readily derived: if

$$w(t) = u(t)v(t) \tag{2.11}$$

then the corresponding Fourier transforms satisfy

$$W(f) = \int_{-\infty}^{\infty} U(f')V(f-f')\,df'. \tag{2.12a}$$

Such an integral is called a convolution. This result may be derived as follows: by substituting the Fourier integrals for $u(t)$ and $v(t)$ we find

$$w(t) = \int_{-\infty}^{\infty} df'\, U(f') \left[\int_{-\infty}^{\infty} df''\, V(f'') \exp\{2\pi j(f'+f'')t\} \right].$$

We then substitute $f = f' + f''$ for f'' in the inner integral, and then interchange the orders of integration:

$$w(t) = \int_{-\infty}^{\infty} df \exp(2\pi jft) \left\{ \int_{-\infty}^{\infty} df'\, U(f')V(f-f') \right\}.$$

(Note that as f'' ranges from $-\infty$ to $+\infty$ in the inner integral with f' fixed, so does the new variable f.) Hence by the uniqueness property the function of f in curly brackets is just $W(f)$, the Fourier transform of $w(t)$, since both give $w(t)$ after an inverse Fourier transform.

The expression (2.12a) looks asymmetric between U and V, but by a simple change of variable to $f'' = f - f'$ it can be altered to

$$W(f) = \int_{-\infty}^{\infty} U(f-f'')V(f'')\,df''. \tag{2.12b}$$

The notation $(U * V)(f)$ or $U(f) * V(f)$ is often used for the convolution on the right-hand side.

The symmetry between t and f points to the following dual theorem, which is also readily verified:

$$\text{If} \quad w(t) = \int_{-\infty}^{\infty} u(t')v(t-t')\,dt' \quad \text{then} \quad W(f) = U(f)V(f) \tag{2.13}$$

Suppose that in the last result we replace $u(t)$ by $u^*(-t)$, so that the Fourier transform $U(f)$ is replaced by $U^*(f)$. Then after substituting $-t'$ for t, we find that:

If $w(t) = \int_{-\infty}^{\infty} u^*(t')v(t+t')\,dt'$ then $W(f) = U^*(f)V(f).$ (2.14)

This integral is called the *correlation function* of $u(t)$ and $v(t)$. The case $v(t) = u(t)$ is especially important, when $w(t)$ is called the *autocorrelation function* of $u(t)$. Thus

If $w(t) = \int_{-\infty}^{\infty} u^*(t')u(t+t')\,dt'$ then $W(f) = |U(f)|^2.$ (2.15)

There is a further specialization. By (2.8b) we find

$$w(0) = \int_{-\infty}^{\infty} W(f)\,df = \int_{-\infty}^{\infty} |U(f)|^2\,df. \qquad (2.16)$$

Another expression for $w(0)$ is found by putting $t = 0$ in (2.15). Equating these expressions we find

$$\int_{-\infty}^{\infty} |U(f)|^2\,df = \int_{-\infty}^{\infty} |u(t)|^2\,dt. \qquad (2.17)$$

This result is known as Parseval's theorem. It is also referred to as the energy theorem since if $u(t)$ is a signal then $|u(t)|^2$ is the instantaneous power and so its time-integral is the energy. Thus if we set $u(t) = \Pi(t)$ (2.3) we find

$$\int_{-\infty}^{\infty} (\operatorname{sinc} f)^2\,df = 1. \qquad (2.18)$$

2.2 The delta function

2.2.1 Definition and properties

The Dirac delta function $\delta(t)$ is a generalized function which may be regarded as the limit of a function which has a tall peak at $t = 0$ with the area under the curve equal to one, and which away from $t = 0$ goes rapidly to zero. Then the peak is made progressively taller and thinner with the area remaining at unity. For example the function $\tau^{-1}\exp(-\pi t^2/\tau^2)$ (for $\tau > 0$) has unit area under its curve (by (2.8a) and Table 2.2) and has a height τ^{-1} and a width of order τ. The delta function is then in a sense made precise by Lighthill (1959) the limit of this function as $\tau \to 0$. In an engineering context it is called a unit impulse at $t = 0$. More generally the function $a\delta(t - t_0)$ represents an impulse of strength a at $t = t_0$. It is remarkable that anything so singular can be manipulated without disaster in the ways described in the previous section. Some useful basic properties are (with $F(t)$ a smooth function):

$$\int_{-\infty}^{\infty} \delta(t - t_0)F(t)\,dt = F(t_0), \qquad (2.19)$$

$$\delta(t-t_0)F(t) = \delta(t-t_0)F(t_0), \tag{2.20}$$

$$\delta(-t) = \delta(t), \tag{2.21}$$

$$\delta(at) = |a|^{-1}\delta(t) \quad (a \neq 0), \tag{2.22}$$

$$\int_a^b \delta(t)\,dt = 1 \quad \text{if} \quad a < 0 \quad \text{and} \quad b > 0. \tag{2.23}$$

The delta function may also be regarded as the derivative of the step function

$$\chi(t) = \begin{cases} 0 & \text{if } t \leq 0, \\ 1 & \text{if } t > 0. \end{cases} \tag{2.24}$$

(The derivatives $\delta'(t)$, $\delta''(t)$ etc. of the delta function may also be defined and used in a straightforward manner. See Exercise 2.8.)

2.2.2 *Fourier transform*

The Fourier transform of $\delta(t - t_0)$ is obtained from (2.19) and (2.1) as $\exp(-2\pi j f t_0)$. Similarly from (2.2) the function of t whose Fourier transform is $V(f) = \delta(f - f_0)$ is $v(t) = \exp(2\pi j f_0 t)$. Thus we may say that the frequency spectrum of a simple harmonic function of frequency f_0 consists of a sharp line at $f = f_0$. Hence from (2.1) we find

$$\delta(f - f_0) = \int_{-\infty}^{\infty} \exp\{-2\pi j(f - f_0)t\}\,dt \tag{2.25}$$

which, in spite of the fact that the integral is not convergent in a normal sense, is a perfectly reasonable relation in the theory of generalized functions. Equation (2.25) is in a sense the limit as $\tau \to \infty$ of

$$\tau \exp\{-\pi \tau^2 (f-f_0)^2\} = \int_{-\infty}^{\infty} \exp(-\pi t^2/\tau^2) \exp\{-2\pi j(f-f_0)t\}\,dt \tag{2.26}$$

(a result readily obtained from (2.7)).

2.2.3 *Filters*

Filters will be considered as carrying out three main tasks:

(a) to separate out the band of frequencies used for communication and to reject everything else;
(b) to reshape pulses to minimize detection errors;
(c) to implement certain definite integrals described later.

However, we start by defining a filter as follows:

A filter is a time-independent system with an output linearly dependent on an input.

THE DELTA FUNCTION

The linearity property means that if the outputs are $v_i(t)$ for a set of inputs $u_i(t)$ ($i = 1, 2, \ldots$) then the output for an input which is the linear combination $a_1 u_1(t) + a_2 u_2(t) + \ldots$ is just $a_1 v_1(t) + a_2 v_2(t) + \ldots$ where $a_1, a_2, \ldots$ are constants. (Although in most cases the inputs and outputs are real it is usually convenient to regard them as complex and specialize to the real case when needed.) We shall extend this idea of linear superposition to the case when the sum over i is replaced by an integral. Thus if the response to an input function $u(\tau, t)$ labelled by a parameter τ is $v(\tau, t)$ then the response to the input

$$\int_{-\infty}^{\infty} \alpha(\tau) u(\tau, t) \, d\tau$$

is the output

$$\int_{-\infty}^{\infty} \alpha(\tau) v(\tau, t) \, d\tau,$$

where $\alpha(\tau)$ is some reasonable function of τ. Naturally the algebraic concept of linearity may be complicated by problems of convergence, but we shall not worry about this any further.

The property of time-independence means simply that if the response to $u(t)$ is $v(t)$, then the response to the shifted input $u(t - \tau)$ is $v(t - \tau)$, where τ has any real value.

2.2.4 Impulse response function

We now introduce the impulse response function $h(t)$ of a filter, which as its name suggests is simply the response to a unit impulse $\delta(t)$ at $t = 0$ as input. One very obvious property of $h(t)$ is

$$h(t) = 0 \quad \text{if} \quad t < 0 \tag{2.27}$$

quite simply because there can be no response before the input. This relation is called the 'causality condition'.

Using $h(t)$ we may readily find an expression for the output corresponding to an arbitrary input $u(t)$. We first note that by time-independence the response to $\delta(t - \tau)$ is $h(t - \tau)$. Then using (2.19) we write the input as a linear combination of time-shifted delta functions labelled by a parameter τ:

$$u(t) = \int_{-\infty}^{\infty} u(\tau) \delta(t - \tau) \, d\tau. \tag{2.28}$$

Hence by linearity the output is

$$v(t) = \int_{-\infty}^{\infty} u(\tau) h(t - \tau) \, d\tau \tag{2.29a}$$

which may also be written as

FOURIER THEORY

$$v(t) = \int_{-\infty}^{\infty} u(t-\tau)h(\tau)\,d\tau \qquad (2.29b)$$

by a change of variable. Evidently the output v is the convolution $u * h$.

This result is so important and yet so simple that it is worth illustrating it by a figure. If the area under the curve of $u(t)$ (Fig. 2.1) is cut up into a myriad of tiny strips such as the one from $t = \tau_i$ to $\tau_i + \Delta\tau_i$, then each strip may

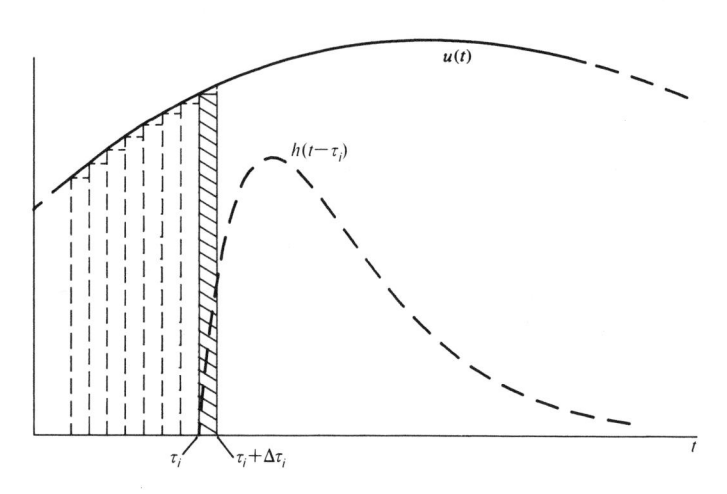

FIG. 2.1 How the output from a filter with impulse-response function $h(t)$ may be regarded as the sum of responses to infinitesimal impulses forming the input $u(t)$.

be regarded as an impulse of strength $u(\tau_i)\Delta\tau_i$, the area of the strip. Hence the response to this particular strip is $h(t - \tau_i)u(\tau_i)\Delta\tau_i$. We then sum over all the strips and replace the sum over i by an integral, in the way that is done to illustrate Riemann integration in elementary calculus texts. In this way we obtain (2.29).

2.2.5 Transfer function

We are now ready to use the methods of Section 2.1. From (2.29b) we find immediately that the response to a simple harmonic input $\exp(2\pi jft)$ is $H(f)\exp(2\pi jft)$, where $H(f)$ is the Fourier transform of $h(t)$:

$$H(f) = \int_{-\infty}^{\infty} h(\tau)\exp(-2\pi jf\tau)\,d\tau. \qquad (2.30)$$

Thus we find the very general and simple result that the response to a unit simple harmonic input is also simple harmonic with an amplitude $H(f)$ which

THE DELTA FUNCTION 23

is the Fourier transform of the impulse response function $h(t)$. We may also take the Fourier transform of the convolution (2.29b) to obtain

$$V(f) = H(f)U(f) \qquad (2.31)$$

where U and V are the respective Fourier transforms of u and v. The quantity $H(f)$ is called the *transfer function* of the filter and may be regarded as a complex gain factor whose magnitude determines the gain at frequency f and whose argument or phase-angle determines the phase-shift. In most cases the input and output of a filter are real so that of course $h(t)$ may also be real and $H(f)$ must obey the reality condition

$$H(-f) = \{H(f)\}^*. \qquad (2.32)$$

The limitations that the causality condition (2.27) imposes on $H(f)$ will be illustrated in due course.

2.2.6 *Some examples of filters*

(a) The output $v(t)$ of a so-called 'RC' filter is related to the input by

$$dv/dt + av = u, \qquad (2.33)$$

where a is a real and positive constant. What are $H(f)$ and $h(t)$? The transfer function $H(f)$ is simple to find. We set $u(t) = \exp(2\pi jft)$, $v(t) = H(f)\exp(2\pi jft)$ in the above equation and after cancelling the factor $\exp(2\pi jft)$ we find

$$H(f) = 1/(a + 2\pi jf). \qquad (2.34)$$

The impulse response function may be obtained from the inverse Fourier transform

$$h(t) = \int_{-\infty}^{\infty} \exp(-2\pi jft)/(a + 2\pi jf) \, df$$

by contour integration or by the use of integral tables.

Another way is to verify that the answer

$$h(t) = \begin{cases} \exp(-at) & \text{for } t > 0, \\ 0 & \text{for } t \leq 0 \end{cases}$$

does have the right Fourier transform. A third argument is as follows: the derivative of a unit step function is a delta function. Therefore at $t = 0$ v must make a step of height 1 so that $dv/dt = \delta(t)$ in (2.33). The finite term $av(t)$ is negligible in comparison with $\delta(t)$ at $t = 0$. Thereafter $h(t)$ is just the exponentially decaying solution of $dv/dt + av = 0$. The magnitude of the transfer function falls off inversely with f at high frequencies which are thus attenuated.

(b) Suppose that $v(t) = u(t) - u(t - \tau)$, a response that may be obtained

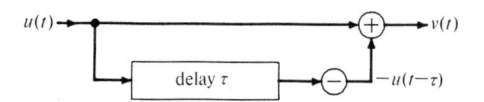

FIG. 2.2 Filter with impulse-response function $\delta(t) - \delta(t-\tau)$.

by the network shown in Fig. 2.2. Then $h(t) = \delta(t) - \delta(t-\tau)$ and $H(f) = 1 - \exp(-2\pi j f \tau)$. If a signal $u(t)$ is band-limited to frequencies below $f_0 = 1/\tau$, that is, $|U(f)|$ vanishes for $|f| > f_0$, then passing the signal through such a filter suppresses both the low-frequency and high-frequency components, including the d.c. component.

(c) A filter much used in later chapters is the *sliding-average* filter where the output v is related to the input u by

$$v(t) = \tau^{-1} \int_{t-\tau}^{t} u(t') dt', \qquad (2.35)$$

with τ a given positive constant. We see that

$$v(t) = \int_{-\infty}^{\infty} u(t') h(t-t') dt'$$

where

$$h(t) = \begin{cases} \tau^{-1} & \text{for } 0 < t < \tau \\ 0 & \text{otherwise.} \end{cases} \qquad (2.36)$$

Thus

$$H(f) = \int_0^{\tau} \tau^{-1} \exp(-2\pi j f t) \, dt = \{1 - \exp(-2\pi j f \tau)\}/(2\pi j f). \qquad (2.37)$$

We may also write

$$H(f) = \exp(-\pi j f \tau) \operatorname{sinc}(f\tau). \qquad (2.38)$$

The transfer function differs from that of the previous example only by the factor $(2\pi j f)^{-1}$, so that this filter is equivalent to the previous filter followed by an 'integrator', a filter with a transfer function equal to $(2\pi j f)^{-1}$, at least in the range of frequencies of interest. (See Section 2.1.2, paragraph (e).)

(d) In some applications we want a good approximation to an ideal *low-pass* filter with a transfer function satisfying

$$|H(f)| = \begin{cases} 1 & \text{for } |f| < f_M, \\ 0 & \text{for } |f| \geq f_M. \end{cases} \qquad (2.39)$$

The cut-off or maximum frequency is f_M. The impulse-response function must be real and in consequence of (2.9) $|H(f)|$ must be symmetric about $f = 0$. Suppose we try to arrange that $H(f) = 1$ if $|f| < f_M$ and zero otherwise. Then by (2.2) and (2.5) the inverse Fourier transform is

$$h(t) = 2f_M \text{ sinc}(2f_M t).$$

This function clearly violates the causality condition (2.27). We can remedy this situation by taking two steps. Firstly $h(t)$ is multiplied by a smooth bell-shaped *windowing* function $s(t)$ which vanishes if $|t| > t_0$ (Oppenheim and Schafer 1975, Section 5.5). We then delay the result by this time t_0. We have a new response function $k(t)$ which does obey the causality condition. The Fourier transform $K(f)$ of $k(t)$ may be qualitatively related to $H(f)$ as follows: the Fourier transform $S(f)$ of $s(t)$ will be a peaked real function of width of order t_0^{-1} centred on $f = 0$. Convoluting this with $H(f)$ will blur the sharp discontinuities of $H(f)$ into falling off over a frequency interval t_0^{-1}. Then the delay will multiply the result by $\exp(2\pi jft_0)$. This situation is illustrated in Fig. 2.3, where f_M has been set equal to 5 and $s(t)$ to the raised-cosine function

$$R(t) = \begin{cases} \{1 + \cos(\pi t)\}/2 & \text{for } |t| < 1, \\ 0 & \text{for } |t| \geq 1, \end{cases}$$

so that t_0 is 1. (The delay has been left out to keep the Fourier transform real for the sake of plotting it.) Thus we find that the causality condition only permits us to set up filters in which the fall-off at the band-edges is over some frequency range t_0^{-1}, and which delay the signal by a time t_0. An almost ideal characteristic may be obtained for large values of t_0 ($t_0 \gg 1/f_M$) but at the cost of a long delay (as well as in practice the cost of an elaborate filter). This delay may be troublesome if the filter is part of a feed-back control loop, but is not

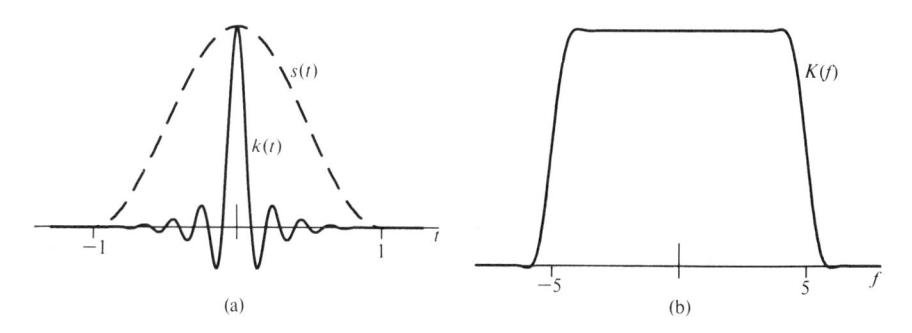

FIG. 2.3 Impulse-response function $k(t)$ and transfer function $K(f)$ of non-ideal low-pass filter. Here $k(t)$ is the product of sinc $(10t)$ with the raised-cosine windowing function $R(t)$ (shown dashed). The function $k(t)$ should be delayed by at least 1 time-unit for the sake of causality, but this delay has been omitted to keep $K(f)$ real.

usually serious in one-way communications. So we shall usually forget about this problem and give our low-pass filters the idealized form of Fig. 2.4(a). Similar remarks may be made about band-pass filters, where the idealized transfer function has the form shown in Fig. 2.4(b).

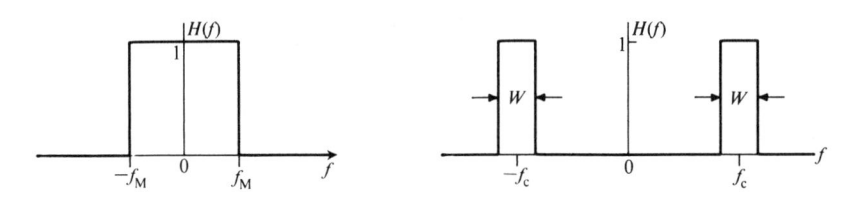

FIG. 2.4 (a) Transfer function for idealized low-pass filter with cut-off frequency f_M; (b) transfer function of idealized band-pass filter with bandwidth W centred on f_c.

2.3 Periodic functions

2.3.1 *Fourier series*

We shall have occasion to deal with periodic functions of time t. We need to discuss such functions for three reasons:

(a) Periodic delta comb functions are introduced in Section 2.4 as a way of deriving the Nyquist theorems, which are basic in the theory of the digital processing of analogue signals.

(b) In Chapter 4 problems arise with certain signals of infinite duration, specifically noise-signals. These problems are side-stepped by forcing the signals to be periodic with a long period which is allowed to tend to infinity at the end.

(c) Periodic functions provide an example of *orthogonal functions* which are used in Chapter 5.

If for some positive constant T a function $\tilde{x}(t)$, defined for all t, satisfies $\tilde{x}(t + T) = \tilde{x}(t)$ for all t, then $\tilde{x}(t)$ is called a *periodic* function of period T. Put simply such a function repeats itself after the time-interval T. (The tilde denotes a periodic function.) The trigonometric functions $\cos(2\pi nt/T)$ and $\sin(2\pi nt/T)$ are periodic functions of period T if n is an integer and it is traditional to regard them as basic or simple. For our purpose the exponential 'simple harmonic' functions $\exp(2\pi jnt/T)$ for integral n are simpler to use. The theory of Fourier series asserts that any function $\tilde{x}(t)$ from a large class of periodic functions of period T may be regarded as a linear combination of these exponential functions (or equivalently of the trigonometric functions)

$$\tilde{x}(t) = \Sigma_n X_n \exp(2\pi jnt/T). \tag{2.40}$$

(Here and from now on the limits on a sum are assumed to be $-\infty$ to $+\infty$

PERIODIC FUNCTIONS

unless otherwise indicated.) The X_n are called the Fourier coefficients and the expansion is called a Fourier series. The class of functions which may be thus expanded includes all the periodic functions of interest to us. The theory is easy to justify for smooth periodic functions and its extension to other functions such as periodic rectangular pulse-trains and impulse trains can also be justified by regarding these as limiting cases of smooth functions.

We proceed to determine the X_n in (2.40) as follows. We use the simple result that for p an integer,

$$T^{-1} \int_{-T/2}^{T/2} \exp(2\pi j pt/T) \, dt = \begin{cases} 1 & \text{if } p = 0, \\ 0 & \text{if } p \neq 0. \end{cases} \quad (2.41)$$

Hence by multiplying (2.40) by $T^{-1} \exp(-2\pi jmt/T)$ where m is any integer and integrating term by term we obtain

$$T^{-1} \int_{-T/2}^{T/2} \tilde{x}(t) \exp(-2\pi jmt/T) \, dt = \Sigma_n X_n T^{-1} \int_{-T/2}^{T/2} \exp\{2\pi j(n-m)t/T\} \, dt. \quad (2.42)$$

Only one term on the right-hand side is non-zero, the one with $n = m$, and hence we find (after interchanging the two sides of this equation)

$$X_m = T^{-1} \int_{-T/2}^{T/2} \tilde{x}(t) \exp(-2\pi jmt/T) \, dt. \quad (2.43)$$

Since the integrand is periodic with period T we may shift the whole range of integration along by changing both limits of integration by the same amount. We may also of course replace the symbol m by n. Thus another version of this result is

$$X_n = T^{-1} \int_0^T \tilde{x}(t) \exp(-2\pi jnt/T) \, dt.$$

We have assumed here that the order of summation and integration can be interchanged. We shall continue to do so, and also make the assumption that if two Fourier series sum to the same function $\tilde{x}(t)$ then the corresponding coefficients are equal. This will be called the 'uniqueness property' of Fourier series. From this we conclude that the equality in (2.40) applies only if the Fourier coefficients are given by (2.43). Nothing else will do. In particular if a Fourier series sums to zero for all t then the Fourier coefficients *must* all be zero.

Example. Let $\tilde{x}(t) = 1/(1 - 2a \cos t + a^2)$ with $0 \leq a < 1$. This function is a smooth periodic function period $T = 2\pi$, which attains a peak value $(1-a)^{-2}$ for t a multiple of 2π and a minimum $(1+a)^{-2}$ halfway in between consecutive peaks. The Fourier coefficients (2.43) are given by

$$X_n = (2\pi)^{-1} \int_{-\pi}^{\pi} \exp(jnt)(1 - 2a\cos t + a^2)^{-1} \, dt = (1-a^2)^{-1} a^{|n|}.$$

(See Gradshtein and Ryzhik 1965, Section 3.613.) As is typical for 'smooth' functions the coefficients fall off with n faster than any power of n, *exponentially* in this case. There are thus no problems caused by poor convergence. (*Note*: a sequence of values s_n will be said to fall off exponentially with n if $|s_n| < A e^{-\alpha n}$ for some fixed positive A and α. The right-hand side can also be written as $A\gamma^n$ with $\gamma = e^{-\alpha} < 1$.)

Example. Let $\tilde{x}(t)$ consist of a string of rectangular pulses of duration τ and height τ^{-1}, repeated with a period $T (T > \tau)$. More precisely let $\tilde{x}(t)$ be a periodic function of period T such that in the range $-\frac{1}{2}T < t < \frac{1}{2}T$ it is given by τ^{-1} if $|t| < \frac{1}{2}\tau$ and zero otherwise. Hence by (2.43) we find that

$$X_n = (\tau T)^{-1} \int_{-\tau/2}^{\tau/2} \exp(-2\pi jnt/T)\, dt$$

$$= \begin{cases} T^{-1} & \text{if } n = 0, \\ T^{-1} \sin(\pi n\tau/T)/(\pi n\tau/T) & \text{if } n \neq 0. \end{cases}$$

In this example the series (2.40) is much more slowly convergent, the terms falling off only inversely with n. The usual way of interpreting an infinite sum as in (2.40) is the limit of N going to infinity of a partial sum

$$s_N = \sum_{n=-N}^{N} X_n \exp(2\pi jnt/T).$$

To show that this is not a very good idea such a partial sum has been plotted in Fig. 2.5(a) (with $\tau/T = 0.4$ and $N = 30$). The 'ears' are caused by overshooting and it can be shown that as N gets very large they become thinner but still maintain their height. This is known as Gibbs' phenomenon. An alternative approach is to multiply each term of the infinite sum by a fairly smooth 'convergence factor' like $\exp(-\pi n^2/N^2)$, so that it converges properly. Then we take the limit $N \to \infty$. (Convergence factors which go to zero fairly smoothly at a finite value N of $|n|$ may also be used to obviate the need for infinite sums.) Figure 2.5(b) shows that in the case $N = 30$ the sum is better behaved, and more as we would expect. The use of a 'convergence factor' is a form of windowing (Section 2.2.6) and is a practical method for getting good answers from slowly convergent Fourier series. It can even be used to make sense of Fourier series which are usually considered divergent, for instance if $X_n = 1$ for all n (see (2.55) below).

Example. Let $\tilde{z}(t) = \tilde{x}(t)\tilde{y}(t)$, where $\tilde{x}(t)$ and $\tilde{y}(t)$ are smooth periodic functions of period T. It is evident that $\tilde{z}(t)$ is also such a function, as the notation has anticipated. We wish to express the Fourier coefficients Z_n of $\tilde{z}(t)$ in terms of X_n and Y_n, the Fourier coefficients of $\tilde{x}(t)$ and $\tilde{y}(t)$ respectively. We find

$$\tilde{z}(t) = \Sigma_{mn} X_m Y_n \exp\{2\pi j(m+n)t/T\} = \Sigma_m X_m [\Sigma_n Y_n \exp\{2\pi j(m+n)/T\}].$$

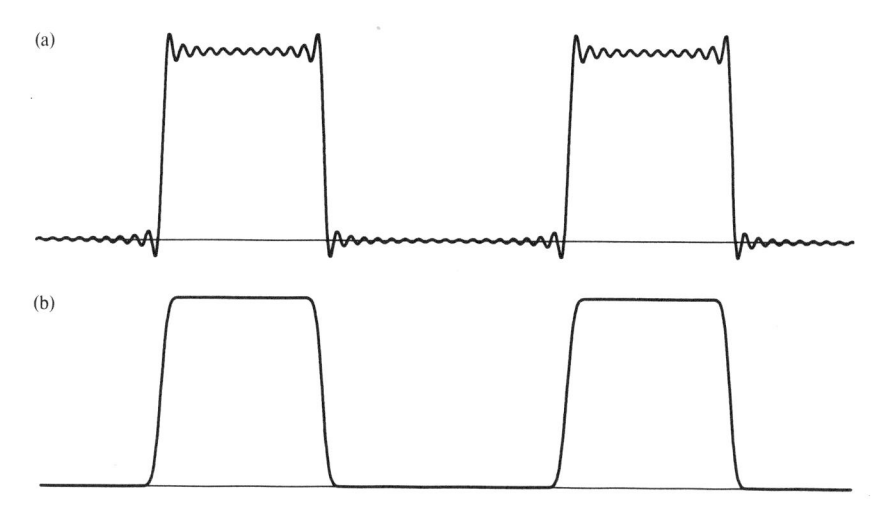

FIG. 2.5 (a) Fourier series for square wave, summed from $n = -30$ to $+30$ (pulse-duration of square wave equals 40 per cent of period); (b) Fourier series for the same square wave with nth term multiplied by $\exp(-\pi n^2/N^2)$ with $N = 30$.

In the inside sum we set $p = m + n$. Then for fixed m we find that as n ranges from $-\infty$ to $+\infty$ so does p. Thus we may write

$$\tilde{z}(t) = \Sigma_m X_m \{\Sigma_p Y_{p-m} \exp(2\pi j p t/T)\}.$$

We rearrange this as

$$\tilde{z}(t) = \Sigma_p \exp(2\pi j p t/T)\{\Sigma_m X_m Y_{p-m}\}.$$

But since $z(t) = \Sigma_p Z_p \exp(2\pi j p t/T)$, we find by the uniqueness property that

$$Z_p = \Sigma_m X_m Y_{p-m}. \tag{2.44}$$

(We may of course replace the letter p by n if we wish.) Note that the Fourier coefficients Z_n have not been derived from (2.43) (although this is not difficult). Note also the device of replacing the sum over n by the sum over p. (This result is analogous to (2.12), although not so useful.)

2.3.2 *A few simple results*

At this stage we list a few simple but useful results analogous to results for Fourier transforms:

(a) There is a 'shifting' theorem that for any fixed τ the Fourier coefficients of the function

$$\tilde{y}(t) = \tilde{x}(t - \tau) \tag{2.45a}$$

are related to the Fourier coefficients X_n of $\tilde{x}(t)$ by

$$Y_n = \exp(-2\pi j nt/T)X_n. \tag{2.45b}$$

For it is evident from (2.40) that

$$\tilde{y}(t) = \tilde{x}(t-\tau) = \Sigma_n X_n \exp(-2\pi j n\tau/T) \exp(2\pi j nt/T),$$

and hence the result follows from the uniqueness property.

(b) We find that if $\tilde{y}(t) = \tilde{x}(-t)$ then $Y_n = X_{-n}$ by replacing t by $-t$ and n by $-n$ in the Fourier series for $\tilde{x}(t)$.

(c) If $\tilde{y}(t) = \tilde{x}^*(t)$ then $Y_n = X^*_{-n}$.

(d) Putting $\tilde{y}(t) = \tilde{x}(t)$ in this result leads to a 'reality condition' for Fourier series, analogous to (2.9):

$$\text{if } \tilde{x}(t) \text{ is real then } X_{-n} = X_n^* \tag{2.46}$$

for all n.

(e) An important result is the 'power theorem' or Parseval's theorem, analogous to (2.17). The average of $|x(t)|^2$ over a period T is given by

$$T^{-1} \int_{-T/2}^{T/2} \tilde{x}^*(t)\tilde{x}(t)\,dt = \Sigma_{mn} X_m^* X_n \left\{ T^{-1} \int_{-T/2}^{T/2} \exp[2\pi j(n-m)t/T]\,dt \right\}.$$

(We use (2.40) twice over, but the name of the subscript is changed in one case to avoid duplication.) Since by (2.41) the term in curly brackets is unity if $m = n$ and zero otherwise, we obtain

$$T^{-1} \int_{-T/2}^{T/2} |\tilde{x}(t)|^2 \,dt = \Sigma_n |X_n|^2. \tag{2.47}$$

If we normalize the signal amplitude of a real signal $x(t)$ so that the instantaneous power is $\{x(t)\}^2$, then the left-hand side is simply the power averaged over a period. A complex function $\tilde{x}(t)$ may be regarded as representing two real signals $\text{Re}\{\tilde{x}(t)\}$ and $\text{Im}\{\tilde{x}(t)\}$, and since the square modulus is the sum of the squares of these quantities it is the sum of the powers of the two signals. Thus we have the useful result that the average power is given by the sum of the square moduli of the Fourier coefficients. It is important to note that the sum extends over negative values of n as well as positive.

2.3.3 Filtering a periodic function

If a periodic function $\tilde{x}(t)$ is put into a filter with impulse-response function $h(t)$, then the output (see 2.29b)

$$\tilde{y}(t) = \int_{-\infty}^{\infty} h(t')\tilde{x}(t-t')\,dt' \tag{2.48}$$

is also periodic. To show this we note that

PERIODIC FUNCTIONS

$$\tilde{y}(t+T) = \int_{-\infty}^{\infty} h(t')\tilde{x}(t-t'+T)\,dt'$$

$$= \int_{-\infty}^{\infty} h(t')\tilde{x}(t-t')\,dt' = \tilde{y}(t).$$

We now relate the Fourier coefficients Y_n of $\tilde{y}(t)$ to those of $\tilde{x}(t)$. If we expand $\tilde{x}(t)$ as a Fourier series by (2.40), then we find that

$$\tilde{y}(t) = \int_{-\infty}^{\infty} h(t')\Sigma_n X_n \exp\{2\pi j n(t-t')/T\}\,dt'$$

$$= \Sigma_n \exp(2\pi jnt/T) X_n \int_{-\infty}^{\infty} h(t') \exp(-2\pi jnt'/T)\,dt'$$

$$= \Sigma_n H(n/T) X_n \exp(2\pi jnt/T),$$

where $H(n/T)$ is the Fourier transform of $h(t)$ evaluated at $f = n/T$. Thus by the uniqueness property we see that the Fourier coefficients Y_n of $y(t)$ are given by

$$Y_n = H(n/T)X_n. \tag{2.49}$$

This result will play a vital role in Chapter 4.

It also explains the behaviour of the truncated Fourier sum shown in Fig. 2.5(a). Truncating the Fourier series for $\tilde{x}(t)$ by dropping terms with $|n| > N$ is equivalent to multiplying each Fourier coefficient X_n by $H(n/T)$ where

$$H(f) = \begin{cases} 1 & \text{if } |f| < f_M, \\ 0 & \text{if } |f| \geq f_M, \end{cases}$$

where f_M is chosen so that $N < f_M T < N + 1$. Hence the new Fourier series sums to the function $\tilde{y}(t)$ which by (2.48) is the convolution of $\tilde{x}(t)$ with the inverse Fourier transform of $H(f)$, given by

$$h(t) = 2f_M \operatorname{sinc}(2f_M t).$$

On the other hand if we choose

$$H(n/T) = \exp(-\pi n^2/N^2)$$

so that

$$H(f) = \exp(-\pi \tau^2 f^2)$$

with $\tau = T/N$, then we are convoluting $\tilde{x}(t)$ with the inverse Fourier transform

$T^{-1}\exp(-\pi t^2/\tau^2)$. This gives rise to a smooth curve where the sharp edges have been blurred (Fig. 2.5b).

2.3.4 Periodically repeated functions

One important application is the following. Suppose that

$$\tilde{x}(t) = \Sigma_k v(t - kT) \tag{2.50}$$

where $v(t)$ is a given function sufficiently localized to make this sum convergent. As usual the sum is from $-\infty$ to $+\infty$. Then $\tilde{x}(t)$ is a periodic function of period T and thus may be expanded as a Fourier series (2.40). Then the Fourier coefficients X_n may be expressed in terms of the Fourier transform $V(f)$ of $v(t)$ as follows:

$$X_n = T^{-1} \int_{-T/2}^{T/2} \Sigma_k v(t-kT) \exp(-2\pi j n t/T) \, dt$$

$$= T^{-1} \Sigma_k \int_{-T/2}^{T/2} v(t-kT) \exp(-2\pi j n t/T) \, dt.$$

In the kth term in this sum we set $t' = t - kT$ and use the fact that

$$\exp(-2\pi j n t/T) = \exp(-2\pi j n t'/T)$$

since $\exp(-2\pi j n k)$ is one. Hence we obtain

$$X_n = T^{-1} \Sigma_k \int_{-kT}^{-(k-1)T} v(t') \exp(-2\pi j n t'/T) \, dt'.$$

We note that the ranges of integration for successive values of k are contiguous, that is they touch each other, and so they may be combined into one range from $-\infty$ to $+\infty$. Thus we find

$$X_n = T^{-1} \int_{-\infty}^{\infty} v(t) \exp(-2\pi j n t/T) \, dt$$

where t' has been replaced by t. Hence by (2.1) we see that

$$X_n = T^{-1} V(n/T). \tag{2.51}$$

2.3.5 The looped wave-guide

For some applications we have to remember that a signal is not just a mathematical concept, but a physical object, an electromagnetic disturbance. It is thus subject to the laws of physics. There are two situations where this has to be taken into account: (a) the study of thermal noise and (b) the study of quantum limitations and the effect of the quantum uncertainty principle. The following

PERIODIC FUNCTIONS 33

idealized physical model will be used. A signal of duration less than T is fed into the input of a long wave-guide (or coaxial cable) along which it propagates without attenuation or dispersion, that is, it does not disappear or change its waveform. (It will be assumed that there is only one mode of propagation, as in a coaxial cable.) The time of propagation from input to output is T. The wave-guide is looped round so that the output is just by the input. After the signal has all been fed in, the input is quickly switched from the signal-source to the output. Thereafter the signal propagates indefinitely round the loop, repeating itself with period T.

The loop is a form of electromagnetic cavity and we now examine the normal modes. They are simple harmonic waves with n cycles distributed round the loop (see Fig. 2.6, where $n = 12$). We need not consider the static case $n = 0$

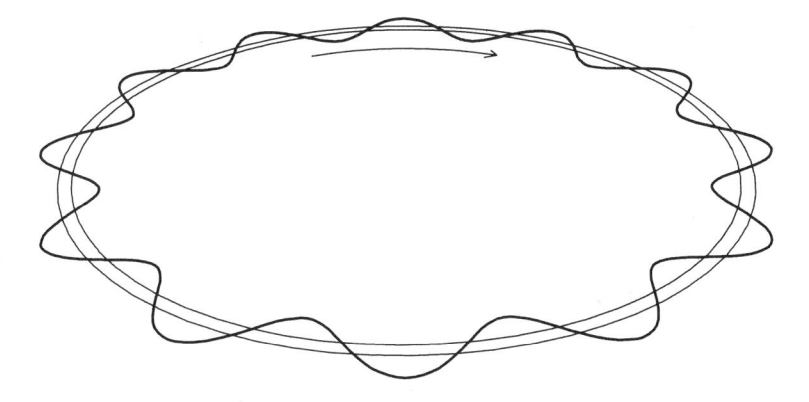

FIG. 2.6 A normal mode in an ideal looped wave-guide, circulating endlessly.

nor those modes which propagate round the loop the wrong way. The signal amplitude at a given point O on the loop due to the nth mode alone being excited is of the form

$$A_n \cos(2\pi nt/T + \theta_n).$$

The frequency is simply n/T because the whole pattern in Fig. 2.6 rotates once every T seconds, and so with n cycles the period of oscillation at O must be T/n. The other quantities A_n and θ_n are respectively the real amplitude and phase of the excitation of this normal mode. The total physical energy is proportional to A_n^2, and we scale our amplitudes so that the energy is just A_n^2.

If several normal modes are excited, then the signal is the sum of the amplitudes of each normal mode and may be written as

$$\tilde{x}(t) = \Sigma_{n=1}^{\infty} A_n \cos(2\pi nt/T + \theta_n).$$

With the definition

$$X_n = \tfrac{1}{2} A_n \exp(j\theta_n),$$

the nth term may be written as

$$X_n \exp(2\pi j nt/T) + X_n^* \exp(-2\pi j nt/T).$$

If, further, we define X_{-n} as X_n^* (for positive n) and $X_0 = 0$, we see that

$$\tilde{x}(t) = \Sigma_{n=-\infty}^{\infty} X_n \exp(2\pi j nt/T).$$

Thus we have a new way of looking at the Fourier series of the real periodic signal in the loop. For every positive value of n the term $X_n \exp(2\pi j nt/T)$ and its complex conjugate $X_{-n} \exp(-2\pi j nt/T)$ add up to give the real amplitude of the nth mode. The magnitude and phase of X_n are respectively the magnitude and phase of the excitation.

The energy of excitation of each mode is readily calculated. Suppose that the nth mode alone is excited, to give the signal $A_n \cos(2\pi nt/T + \theta_n)$. The square of this is the instantaneous power and so the energy is obtained by integrating the square over a complete period T to give the energy of the nth mode as $\frac{1}{2}A_n^2 T$ or as $2TX_n^* X_n$.

2.4 Delta combs

2.4.1 *Fourier transform of a periodic function*

The Fourier transform $X(f)$ of a periodic function $\tilde{x}(t)$ of period T as calculated by (2.1) diverges, but is readily expressible in terms of delta functions. Since the Fourier transform of $\exp(2\pi j nt/T)$ is $\delta(f - n/T)$, by (2.1) and (2.25) we find that by a term-by-term Fourier transform of the Fourier series (2.40) for $\tilde{x}(t)$,

$$\tilde{x}(t) = \Sigma_n X_n \exp(2\pi j nt/T),$$

we obtain the Fourier transform of $\tilde{x}(t)$:

$$X(f) = \Sigma_n X_n \delta(f - n/T). \tag{2.52}$$

Thus the frequency spectrum of $\tilde{x}(t)$ is a series of delta functions at multiples of the basic frequency $1/T$. Such a function is known as a *delta comb*.

2.4.2 *The periodic comb*

The periodic function

$$c_T(t) = \Sigma_k \delta(t - kT) \tag{2.53}$$

is called a periodic delta comb with period T. The subscript T on c is usually left out. The special case with $T = 1$ is called the 'sha' function after the Russian character Ш

DELTA COMBS

$$\text{III}(t) = \Sigma_k \delta(t-k). \tag{2.54}$$

In particular the scaling rule (2.22) gives

$$c(t) = T^{-1}\text{III}(t/T).$$

The Fourier coefficients of $c(t)$ are given by (2.43) and (2.19) as

$$C_n = T^{-1}\int_{-T/2}^{T/2} \delta(t)\exp(-2\pi jnt/T)\,dt = T^{-1}.$$

(Although the range of integration is limited it may be extended to $\pm\infty$ since $\delta(t) = 0$ for $|t| \geqslant T/2$.) Thus by (2.40) we find

$$c(t) = T^{-1}\Sigma_n \exp(2\pi jnt/T). \tag{2.55}$$

From (2.52) and (2.22) we find that the Fourier transform $C(f)$ of $c(t)$ is given by

$$C(f) = T^{-1}\Sigma_n \delta(f - n/T) = \text{III}(Tf). \tag{2.56}$$

Thus in particular we find by setting $T = 1$ that the Fourier transform of $\text{III}(t)$ is just $\text{III}(f)$.

We also note the following results:

$$x(t)c(t) = x(t)\Sigma_k \delta(t - kT) = \Sigma_k x(kT)\delta(t - kT) \tag{2.57}$$

$$x(t) * c(t) = \int_{-\infty}^{\infty} x(t-t')\Sigma_k \delta(t' - kT)\,dt' = \Sigma_k x(t - kT), \tag{2.58}$$

$$X(f)C(f) = T^{-1}\Sigma_n X(n/T)\delta(f - n/T), \tag{2.59}$$

and

$$X(f) * C(f) = T^{-1}\Sigma_n X(f - n/T). \tag{2.60}$$

The periodic delta comb may be used for a quick derivation of results that may be proved in other ways:

(a) The result (2.51) of Section 2.3.4 may be obtained from (2.50) as follows: we note that by (2.50) and (2.58) $\tilde{x}(t)$ is the convolution of $c(t)$ and $v(t)$. Hence the Fourier transform $X(f)$ is given by

$$X(f) = C(f)V(f) = T^{-1}\Sigma_n \delta(f - n/T)V(f) = T^{-1}\Sigma_n \delta(f - n/T)V(n/T).$$

Thus comparing with (2.52) we see that the Fourier coefficients are

$$X_n = T^{-1}V(n/T).$$

(b) If the Fourier transform of a well-behaved function $u(t)$ satisfies

$$\Sigma_n U(f - n/T) = 1 \tag{2.61a}$$

for all f then it is found that

$$u(kT) = 0 \quad \text{for all non-zero integral values of } k. \quad (2.61\text{b})$$

An example of a function $U(f)$ with this property is shown in Fig. 2.7. We find

$$1 = \Sigma_n U(f - n/T) = TU(f) * C(f)$$

by (2.60). So by a Fourier transform we find

$$\delta(t) = Tc(t)u(t) = T\Sigma_k \delta(t - kT)u(t) = T\Sigma_k \delta(t - kT)u(kT).$$

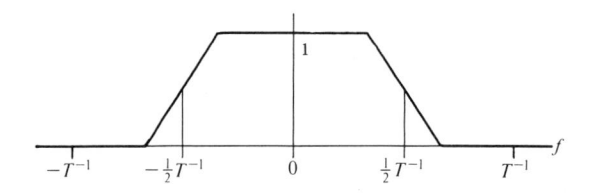

FIG. 2.7 A transfer function with the Nyquist property.

The right-hand side is a delta comb but it is evident from the left-hand side that all the teeth are missing except the one at $t = 0$. Hence we see that $u(kT) = 0$ for $k \neq 0$ and $u(0) = T^{-1}$. (This is one of the Nyquist theorems used in the theory of pulsed signalling—Section 3.3.1. The more famous 'sampling' theorem will be discussed in Section 3.2.5.)

For further reading: Bracewell's monograph (1978) is highly recommended for readers who are not familiar with Fourier theory, and who would thus be helped by a development of the material of this chapter in a manner more leisurely than that which is possible here.

Exercises

2.1. Confirm the second, third, fourth, and fifth results of Table 2.1.

2.2. By separating

$$\int\int_{-\infty}^{\infty} \exp\{-(x^2 + y^2)\} \, dx \, dy$$

in cartesians, and by evaluating it in polar coordinates, show that

$$\int_{-\infty}^{\infty} \exp(-x^2) \, dx = \sqrt{\pi}.$$

2.3. Prove that the Fourier transform of $v(t) = \exp(-\pi t^2)$ satisfies

$$V'(f) = -2\pi f V(f)$$

by differentiating (2.1) under the integral sign and then integrating by parts. Hence show that $V(f)$ is $C\exp(-\pi f^2)$ and find C from the result of Exercise 2.2.

2.4. Find the Fourier transform of

$$\text{Re}[\exp(-\pi t^2/\tau^2)\exp\{2\pi j(f_0 t + \alpha t^2/2)\}],$$

with τ, α, and f_0 real. Why is such a function given the name 'chirp function'?

2.5. Prove (a) that $t^{-n}e^{-1/t}$ tends to zero as t tends to zero from the positive side, (b) that every derivative of $e^{-1/t}$ has the form of a polynomial in t^{-1} multiplied by $e^{-1/t}$, (c) and hence that every derivative of $e^{-1/t}$ tends to zero as t tends to zero from the positive side. Hence show that the function given by

$$F(t) = e^{-1/t} \text{ for } t > 0, \quad F(t) = 0 \text{ for } t \leqslant 0$$

is a smooth function, and that $F(t+1)F(1-t)$ is a good function. Show that this function is $\exp\{-2/(1-t^2)\}$ for $|t| < 1$, and 0 for $|t| \geqslant 1$.

2.6. Use the reality condition (2.9) to show that (2.2) may be written as

$$v(t) = \int_0^\infty df\{2\text{Re}\,V(f)\}\cos(2\pi ft) - \int_0^\infty df\{2\text{Im}\,V(f)\}\sin(2\pi ft)$$

provided $v(t)$ is real.

2.7. Find the convolution of the two Gaussian functions $\exp(-at^2)$ and $\exp(-bt^2)$, with a and b real and positive.

2.8. Show that $p'(t) = \delta(t) - \delta(t-1)$ where $p(t)$ is the pulse function given by

$$p(t) = 1 \text{ if } 0 < t < 1, \quad p(t) = 0 \text{ otherwise}.$$

Sketch the graph of $g(t) = tp(t)$ and show that

$$g'(t) = p(t) - \delta(t-1) \quad \text{and} \quad g''(t) = \delta(t) - \delta(t-1) - \delta'(t-1).$$

(Note that it is possible to have a delta function and its derivative 'at the same place' without one swamping the other.) Find the Fourier transform of $g''(t)$ and hence show that the Fourier transform $G(f)$ of $g(t)$ is

$$(2\pi jf)^{-2}(1 - e^{-2\pi jf} - 2\pi jfe^{-2\pi jf}).$$

Show that this function is well-behaved as f tends to 0. Verify this result directly from (2.1) by integrating by parts.

2.9. Find the Fourier transform of the triangular pulse given by

$$g(t) = 1 - |t|/T \text{ for } |t| < T, \quad g(t) = 0 \text{ otherwise}$$

(a) by differentiation to δ-functions as in Exercise 2.8, (b) as a convolution of a rectangular pulse with itself, (c) directly from (2.1) by integration by parts.

2.10. Derive (2.26) from (2.7).

2.11. Show that if an impulse function is symmetric about a time t_0, that is it satisfies $h(t_0 - t) = h(t_0 + t)$ for all t, then $H(f) = H'(f)\exp(-2\pi j f t_0)$ with $H'(f)$ real. (Such a transfer function is said to possess the property of 'linear phase'.)

2.12. Illustrate the impulse-response function of a nearly ideal band-pass filter.

2.13. Derive (2.41).

2.14. Show that if $\Sigma_k u(t - kT) = 1$ for all t and if $\check{x}(t)$ is a periodic function of period T then

$$\int_{-T/2}^{T/2} \check{x}(t)\,dt = \int_{-\infty}^{\infty} \check{x}(t)u(t)\,dt.$$

2.15. What are the conditions on the Fourier coefficients of a periodic function $\check{x}(t)$ if (a) $\check{x}(t)$ is real and symmetric, that is, $\check{x}(-t) = \check{x}(t)$? (b) $\check{x}(t)$ is real and antisymmetric, that is, $\check{x}(-t) = -\check{x}(t)$? Give an example of a periodic function of each type.

2.16. Verify that the relation (2.51) applies for a periodically repeated rectangular pulse.

2.17. Show how the Fourier series for $\check{x}(t)$ given by (2.50) goes over into a Fourier integral as $T \to \infty$. Use this result to produce a plausibility argument for the Fourier inversion theorem of Section 2.1.1.

2.18. Show that the function $U(f)$ in Fig. 2.7 does have the property claimed (2.61a) and that it is the convolution of two rectangular shapes. Hence find $u(t)$ and show that $u(kT) = 0$ for $k \neq 0$.

2.19. Show that

$$\Sigma_{k=-\infty}^{\infty} \exp\{-\pi(t - kT)^2\} = T^{-1}\Sigma_{n=-\infty}^{\infty} \exp(-\pi n^2/T^2)\exp(-2\pi j n t/T).$$

Which series converges more quickly if $T \gg 1$?

2.20. With $\alpha = \exp(2\pi j/N)$ show (by summing a geometric progression or otherwise) that

$$\Sigma_{k=0}^{N-1} \alpha^{lk} = 0 \text{ if } l \text{ is not a multiple of } N,$$

$$= N \text{ if } l \text{ is a multiple of } N.$$

Hence, show that if

$$X_n = \Sigma_{k=0}^{N-1} x_k \alpha^{kn} \quad (n = 0, 1, \ldots N - 1)$$

then

$$x_l = N^{-1} \Sigma_{m=0}^{N-1} X_m \alpha^{-lm} \quad (l = 0, 1, \ldots N - 1).$$

(The list of values X_n is called the 'discrete Fourier transform' of x_k.)

3. Digital signalling methods

In this chapter methods of sending digital signals are discussed. The advantages of digital methods over analogue methods have already been described in Section 1.1.3. Digital signals are selected from a discrete set of possibilities, well separated in some sense, whereas analogue signals are selected from a continuous range. Hence small perturbations of digital signals caused by the channel can be recognized and corrected, whereas analogue signals are left at their perturbed values, with no possibility of correction.

In Section 3.1 we discuss how information is measured quantitatively. The treatment is at a superficial level, which is all that is necessary for this book. Curiously enough information theory at a rigorous level does not have much relevance for elementary coding theory, and several texts on coding theory hardly mention it, or discuss it outside the main line of development. The subject of data-compression (otherwise known as source-coding or bit-rate reduction) is also discussed only briefly. In fact it is a huge and important topic, but again it is not very relevant for our theme. (See, for example, Davisson and Gray 1976 or IEEE 1982*a*.)

Digital-pulse signalling is introduced in Section 3.2. A system where the pulses are sent directly down the signalling wire without any form of carrier-modulation is called base-band signalling, because the range of frequencies used extends down to zero or near zero. The problems of obtaining reliability are discussed qualitatively, since as yet there is no quantitative model for the noise on the channel. Then the discussion moves on to how analogue signals are converted to digital with negligible loss of fidelity. This is done in two stages, periodic sampling followed by quantization. At the end of this section the performance of digital and analogue repeaters are contrasted on a long-distance link where the signal has to be boosted at intervals, mostly to illustrate the robustness of digital methods.

The problem of controlling the bandwidths of digital signalling are discussed in Section 3.3. The bandwidth normally must be restricted, especially if the signals are sent by radio. Unfortunately this can lead to a form of error-producing distortion known as inter-symbol interference (ISI), because the necessary filtering makes pulses drag out so that they interfere with subsequent pulses.

Digital signals can be used to modulate a carrier by amplitude modulation, just like analogue sound and vision signals. However there are various other possibilities, some of which are explored in Section 3.4 after an introduction to the theory of carrier-modulation. The analysis in terms of phasors and signal-constellations will be developed into the concept of a signal-space in Chapter 5.

3.1 The concept of information

3.1.1 *Information and information rates*

We need a quantitative and convenient definition of how much information is sent when a message is transmitted over a link. For our purposes the following rather rough definition is adequate: suppose that in a particular interval of time T the link can transmit one of M possible messages. Thus if the link is a teleprinter link which can send any string of n characters from an alphabet of size q in that period, then M is equal to q^n, there being that many n-character strings. (For reasonable values of n this is a huge number. See Exercise 3.1.) If the possible messages are restricted in some way, then of course there are fewer than q^n. For instance the messages might always be English text, or FORTRAN programs. Once we have decided on the value of M we say that the information transmitted is given by

$$I = \log_2 M. \tag{3.1}$$

A logarithmic measure is used for the following reason: suppose that the link is used twice, with M possible messages in the first interval and N in the second. Then in effect one composite message out of MN possibilities is chosen, so that the information sent over both intervals is $\log(MN)$. Since this is $\log M + \log N$ the total amount of information sent is the sum of the amounts in each interval. Any other definition of the amount of information as a function of M does not have this additive property (Exercise 3.2).

The choice 2 for the base of the logarithm is indicated by assigning a unit, the *bit*, to the measure of information. (Sometimes the logarithm in (3.1) is replaced by a natural logarithm, in which case the unit is called a *nat*.) Since

$$\log_2 M = \ln M / \ln 2 \tag{3.2}$$

(a useful formula for computing logarithms to the base 2) it follows that

$$I \ln 2 = \ln M, \tag{3.3}$$

a formula that expresses I in terms of the natural logarithm of M. This is a convenient result if M is given by a theoretical approximation involving the exponential function.

We now suppose that the M messages have to be represented in a binary representation, for instance so that the chosen message can be represented in a computer. One method of doing this, not a very practical one, is to number the messages from 0 to $M - 1$ so that they can be represented by an integer in this range. The number of binary digits needed to store an arbitrary integer in this range is the least integer not less than $\log_2 M$. Thus with $M = 11$ we need 4 binary digits, since $\log_2 11 = \ln 11/\ln 2 = 3.4594$. For large values of M this rounding up makes a very small fractional difference and we shall state that $\log_2 M$ binary digits are needed without worrying about the slight inaccuracy, or the fact that this logarithm is not usually an integer. Thus the number of binary digits needed is equal to the amount of information $I = \log_2 M$ in bits. It is perhaps unfortunate that binary digits are also called bits, since we are then using the same name for two logically different but closely related concepts. If necessary we shall distinguish the concepts by talking about bits of information or binary digits as the situation requires.

Suppose now each message is represented by a string of fixed length l made up out of characters from an alphabet of size q. Since there are q^l strings of length l we need $q^l \geqslant M$ in order to assign a unique character-string to each message. Thus at least $l/(\log_2 q)$ characters are needed to represent the message, and each character is said to convey $\log_2 q$ bits of information. If the characters are simply 0 and 1, in an alphabet of size 2, then each character conveys 1 bit of information. If they are from the 26-letter 'Roman' alphabet of 26 capital letters they each convey $\log_2 26 = 4.700$ bits, and if they are hexadecimal digits from the hexadecimal character set 0, 1, 2, 3, 4, 5, 6, 7, 8, 9, A, B, C, D, E, F they convey 4 bits of information.

A communication system is not usually designed to send a single message, but to be used continuously. Thus the rate of transmission of information is a concept of greater importance for us than the amount of information. It is usually calculated as the amount of information sent over a long time T, divided by T. Thus if n characters per second are being sent the rate of transmission of information is $n \log_2 q$ bits s^{-1}, where q is the size of the alphabet.

3.1.2 Data compression

It is often possible to reduce the bit-rate, the number of bits sent each second, without degrading the link significantly. Such a reduction is known as 'data-compression' or 'bit-rate reduction'. For instance in typical English the character 'e' is much more commonly used than 'z'. So rather than use the ASCII code we might use some other code where the number of bits assigned to 'e' is rather less than the number of bits assigned to 'z'. We might go further and use a code where commonly used words like 'the' and even phrases like 'Happy Christmas!' are given special binary strings. Perhaps the best-known example of such a code is the Morse code where the codeword for the most frequent character E is a single dot whereas for Z it is — — · ·. The Huffman codes are the result of a systematic development of this idea (Hamming 1980).

We follow this point up a little further. It is found that the number M_N of sensible English texts of a length around N letters (to within a small specified percentage) is of order r^N; that is, it varies exponentially with N. The value of r is rather less than the number of letters in the alphabet. In principle to each of these messages could be assigned a number less than M_N which would be represented by about $\log_2 (r^N) = N \log_2 r$ bits. So the message could be squeezed down to $\log_2 r$ bits/character rather than, say, the 7 bits/character of the ASCII code. Estimates for $\log_2 r$ vary from 1.0 to 1.5 bits/character (Beker and Piper 1982, Section 3.6.3). From a practical point of view it means that English texts may be considerably compressed, even if not down to 1.5 bits/character. There are disadvantages in using such data-compression:

(a) The implementation may be costly.

(b) The system may be worse than useless if the text to be sent has unexpected characteristics. Thus the Morse code is very inefficient in transmitting text in a language where Z is the commonest letter and E the rarest.

(c) The capability for detecting and correcting errors is reduced. One can usually correct misprints in conventional English text. This shows that ordinary language has an informal but nevertheless powerful error-correcting capability. This is lost to some extent if data-compression is used. To take an extreme example: suppose we send texts using the method described above, where every text of length around N is assigned a number. Suppose, moreover, a single bit is received in error at the more significant end of this number. The value of the number is changed by a large amount and the corresponding text is probably totally unrelated to what was actually sent. Moreover, since the incorrect text is a sensible English text it might be hard to tell that there is anything wrong. This illustrates the important point that redundancy in a message is not necessarily wasteful because it may be needed for the correction and even the detection of errors.

With this form of data-compression the original text can be regenerated faithfully and completely. Another way of achieving data-compression is by removing useless information from the message, or what is believed to be useless information. It causes an irreversible loss of fidelity since the original message can no longer be reconstituted, but if it is well used the loss is not serious. Here are some examples:

(a) The result of an experimental measurement may be a value like 0.47947135 but it is recorded to two significant figures as 0.48 since the measurement is only accurate to this level of precision.

(b) The older telegraph systems and the old computer line-printers used only the capital letters of the alphabet (the Roman alphabet).

(c) Television and cinema systems use 24 to 30 frames per second, because the human eye cannot respond any faster. There is no point in sending, say, 100 frames per second.

(d) The human ear is incapable of responding to sounds at frequencies higher than 20 kHz. Therefore such frequencies are filtered out from sound-signals to be broadcast or recorded.

(e) Most of the intelligibility of speech is conveyed by frequencies in the range 0.3 to 3.5 kHz. Telephone circuits filter out frequencies not in this range.

In most of our discussions of signalling we shall assume that if there is any data-compression needed it has been carried out before the signal or message is presented to the input of the communication system.

The possibility of using data-compression highlights a problem which was skipped over in the last section, about how we decide which messages are possible. Let us consider messages of length around N. If we take the optimistic point of view that the set consists only of all the possible English messages of length around N, then the amount of information is $N \log_2 r$ where r is the quantity discussed above. However, we prefer to take the point of view that any string of characters is to be considered as likely as any other. Thus (with an alphabet of size q) there are of order q^N possible messages. Hence the amount of information to be transmitted is the larger value $N \log_2 q$. In other words we should be prepared to send that number of bits in communicating the message.

Finally, it should be noted that there is a mathematical theory of information called information theory, which works by assigning a probability to each message. Its techniques are not needed for this book. The interested reader is referred to Hamming (1980) for a very readable introduction.

3.2 Baseband signalling

3.2.1 *Baseband binary and multilevel pulse signalling*

An obvious way of conveying binary information along a wire is to denote 0 by an absence of signal-voltage, and 1 by the presence of a voltage at a specified level V. This method is known as 'on-off keying' (OOK), and is illustrated in Fig. 3.1(a).

An alternative is a *polar* signalling system, with two levels of equal magnitude but opposite sign. The following convention will be used:

$$\text{bit } 0 \leftrightarrow +\text{ve level}, \quad \text{bit } 1 \leftrightarrow -\text{ve level}. \tag{3.4}$$

The ensuing signal is shown in Fig. 3.1(b). In order to compare this signalling system with OOK we choose the levels as $\pm \frac{1}{2} V$. The average power is $(\frac{1}{2} V)^2 = V^2/4$, that is one half that of the previous method, where with an equal number of 0s and 1s the average power is $\frac{1}{2} V^2$. None the less the spacing of the two levels is still V, and so the sensitivity to added interference is the same.

If the interference-level is low enough it is better to use more than two levels for the pulse heights. With a choice of four levels each pulse represents a symbol from an alphabet with $q = 4$ symbols. If the bits are combined according to the scheme $00 \to 0$, $01 \to 1$, $10 \to 2$, $11 \to 3$, then the original message 11010010 is then represented by 3102 as in Fig. 3.1(c). The rate of transmission of information is doubled, since each pulse is worth $\log_2 4 = 2$ bits. The cost is a greater susceptibility to error caused by interference, since for a given average power the levels are more closely spaced.

The receiver has to measure the signal amplitudes against the background of

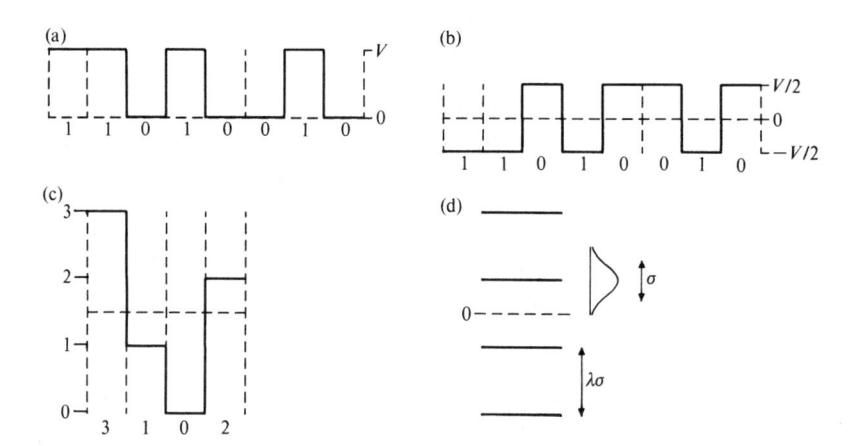

FIG. 3.1 (a) On–Off Keying (OOK) pulsed signal; (b) polar pulsed signal; (c) four-level amplitude shift keying (4-ASK); (d) uncertainty caused by noise in the measured height of a pulse.

interference or *noise* as it is usually called. Rather than making instantaneous measurements it is better to take an average of the signal-level over the duration of each pulse so that the noise-fluctuations tend to cancel out. This average may be found by resetting an integrator to zero at $t = 0$ and then letting it accumulate the time-integral of the received signal up to T_b, when the integrated value is delivered to the output after division by T_b. Then the integrator is reset to zero in anticipation of the next pulse. This process will be called 'integrate and dump'.

3.2.2 Information rate for 'moderate reliability'

How many levels should be used? It will be assumed that the system is constrained to operate at a given mean-power level, and that it is subject to a certain noise-level which gives a random scatter to each pulse-height with a spread of σ (Fig. 3.1(d)). This may be pictured as being caused by the addition of a random noise-signal with an average power $N = \sigma^2$. So the signalling levels should be spaced by $\lambda\sigma$, where λ is a chosen parameter of order unity. With a choice of n levels the largest pulse has an amplitude of $\lambda\sigma \cdot \frac{1}{2}(n-1)$. If all levels are equally likely the average signal power S is then

$$n^{-1}\lambda^2\sigma^2 \sum_{j=-(n-1)/2}^{(n-1)/2}(j^2).$$

By the use of induction (Exercise 3.6) it is found that the sum is $n(n^2-1)/12$. Thus the ratio S/N is equal to $(n^2-1)/\kappa$ with $\kappa = 12/\lambda^2$, a constant of order unity. Hence n is roughly given by $n = \sqrt{(\kappa S/N)}$ for large values of S/N. The information per pulse is $\log_2 n$ and hence the information-rate is roughly $\frac{1}{2}T_b^{-1}\log_2(\kappa S/N)$, where T_b is the pulse-duration, so that T_b^{-1} is the pulse-rate. This is a rough but useful criterion for the rate of transmission of information with 'moderate reliability'.[†]

A similar formula was derived by Shannon (1948) for the 'capacity' of a noisy link or 'channel' (Section 5.3), but in this case the much more powerful result was derived that with suitable coding it is possible to send information at any rate below capacity with arbitrarily high reliability.

3.2.3 Gray codes

In multilevel signalling with additive noise the most likely error is one level being confused with an adjacent level. Thus in a system with 16 levels 0 to 15 an error in detecting level 8 usually gives level 7 or 9 rather than 15 or 0. If these level-numbers correspond to 4-bit strings in the obvious way as binary representations then such an error may change several bits. Thus reading 7 for 8 changes the correct output string 1000 to 0111 so that all four bits are wrong. Fortunately it is possible to code the levels so that such an error changes only one bit. Such a code is known as a Gray code (Schwartz 1980, Chapter 3).

Level 0 is coded as 0, leading zeros are not shown, and then 1 is coded as 1. Then we reflect this pattern in the horizontal line A with a 1 put in the next column (Fig. 3.2.(a)). After 11 and 10 we reflect the whole pattern in the line B, but again with 1s in the third column from the right. Then we reflect the whole pattern in the line C, with 1s in the fourth column. An encoding algorithm which avoids the use of a 'look-up' table is shown in Fig. 3.2(b). To find the code for any level n we represent n as a binary number $b_3 b_2 b_1 b_0$, shift it right, losing the right-hand bit, and perform a 'no-carry' binary addition (i.e. $1 + 1 = 0$) with the unshifted version. The inverse process is shown in Fig. 3.2(c). The proof that this algorithm works is not as simple as its description!

Unfortunately a single error in a Gray code does not necessarily give an adjacent level. This is because there are only two adjacent levels at most, but several possibilities for single-bit errors, four for the code illustrated in Fig. 3.2(a).

3.2.4 Codes and signals

What do the expressions 'code', 'signal', and 'codeword' signify? We suppose that one of a set of M messages is to be sent. It is usually best to regard these messages

---

[†] *Note*: The symbol T_b will be used to denote the duration of a pulse or the pulse-repetition interval, so that T_b^{-1} is the rate at which pulses are sent. The symbol T_W will denote the duration of a signal or codeword which is a train of several pulses or which has a complicated structure. Long periods of time, such as those over which periodicity is assumed for carrying out expansions in Fourier series, will be denoted simply by T.

46 DIGITAL SIGNALLING METHODS

```
         0
         1
        11
   A ───
        11
        10
   B ═══
        110
        111
        101
        100
   C ───
        1100
        1101
        1111
        1110
        1010
        1011
        1001
        1000
```

(a)

(b) [encoding algorithm diagram: $b_3 b_2 b_1 b_0 = 1\,1\,0\,1 \rightarrow c_3 c_2 c_1 c_0 = 1\,0\,1\,1$]

(c) [decoding algorithm diagram: $c_3 c_2 c_1 c_0 = 1\,0\,1\,1 \rightarrow b_3 b_2 b_1 b_0 = 1\,1\,0\,1$]

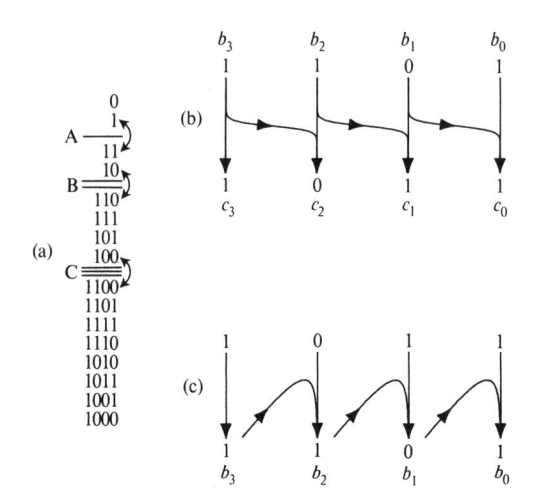

FIG. 3.2 Four-bit Gray code: (a) code table (b) encoding algorithm; (c) decoding algorithm.

as simple objects, say characters in an alphabet rather than whole books. We represent each message by a codeword, a pattern of some sort. Thus in the Morse code each letter is represented by a pattern of dots and dashes and in the ASCII teleprinter code by a pattern of binary digits. The set of all the codewords is called a *code*; this name is also used for the table (code-table) assigning a codeword to each message. (The definition of a code as a collection of codewords without any indication of which each codeword stands for may seem a little odd. It will be seen however that the error-correcting properties of a code depend mostly on the difference between the codewords themselves, and not on what they stand for.) A signal is a physical realization of a codeword so that the message can be sent. It may be a time-dependent voltage for instance. The 'signal set' is the set of all signals corresponding to the messages in the message set, and may itself be regarded as a code if that is convenient, with each signal as a codeword. At the transmitter the message to be sent usually goes through two stages. First there is an *encoder*, which 'looks up' the codeword corresponding to the message to be sent. The encoder then passes on the codeword to a signal generator or *modulator*, which produces the signal.

We may illustrate these ideas by means of the Gray code just described. The message is a 4-bit string $b_3 b_2 b_1 b_0$, one of a set of $M = 16$ possibilities. The encoder for the Gray code performs the operation shown in Fig. 3.2(b), which is equivalent to using a look-up table, and it produces the corresponding codeword $c_3 c_2 c_1 c_0$. The modulator then uses this string as the binary representation of an integer to set the height of the pulse to be sent down the line. This height will be one of $M = 16$ possibilities h_i ($i = 0, \ldots, M - 1$), say.

At the receiving end the process is more complicated, since the pulse-height may have been perturbed by interference on the channel, and the received value will not be exactly equal to the transmitted value. The first job is to measure the pulse-height; this is carried out by the demodulator, which would for instance average the received signal over the duration of each pulse. The second job is to decide from the measured pulse-height r what was the most likely signal to have been sent. If the noise is additive, then a good *decision procedure* is finding which transmitted pulse-height h_i is nearest the received value r, in other words finding which h_i minimizes $|r - h_i|$. After that comes the third relatively trivial task of finding the corresponding message. The second and third tasks are carried out by the *decoder*.

3.2.5 Nyquist sampling of analogue signals

An analogue signal $u(t)$ takes on a value from a continuous range at each moment in time. In order to transmit the signal digitally it has first to be sampled at discrete moments in time and secondly it has to be *quantized*, each sampled value being rounded to a value represented by a specified number of bits. We examine each stage of this process in turn. To encourage us in the first stage there is a remarkable theorem called the Nyquist sampling theorem which states that if a real signal $u(t)$, bandlimited to frequencies below some value f_{max} in magnitude, is sampled at a rate T_b^{-1}, then provided that $T_b^{-1} > 2 f_{max}$ the original signal may be faithfully reconstructed from the samples. This is shown as follows:

The process of sampling and reconstruction is envisaged as taking place in three stages (Fig. 3.3(a)). The sampler produces the sampled values

$$u_k = u(kT_b)$$

from the input signal (Fig. 3.3(b)). These values are sent over an ideal link. At the receiving end an impulse generator produces the signal (Fig. 3.3(c))

$$w(t) = \Sigma_k u_k \delta(t - kT_b).$$

Then a filter produces a smoothed output (Fig. 3.3(e)) by convoluting $w(t)$ with the filter-response function $h(t)$ (Fig. 3.3(d)):

$$v(t) = h(t) * w(t) = \Sigma_k u_k h(t - kT_b).$$

It is evident from this equation that the reconstruction is possible without using the impulse train $w(t)$. However, from the point of view of the theory it is simpler to suppose that the signal is reconstructed in the manner described. (This is a situation that sometimes arises in the analysis of a signal-processing technique. There are two methods for achieving the same end, but whereas one is theoretically easier to analyse, the other is simpler to implement in practice. None the less the theoretical results derived for one method must apply to the other.)

48   DIGITAL SIGNALLING METHODS

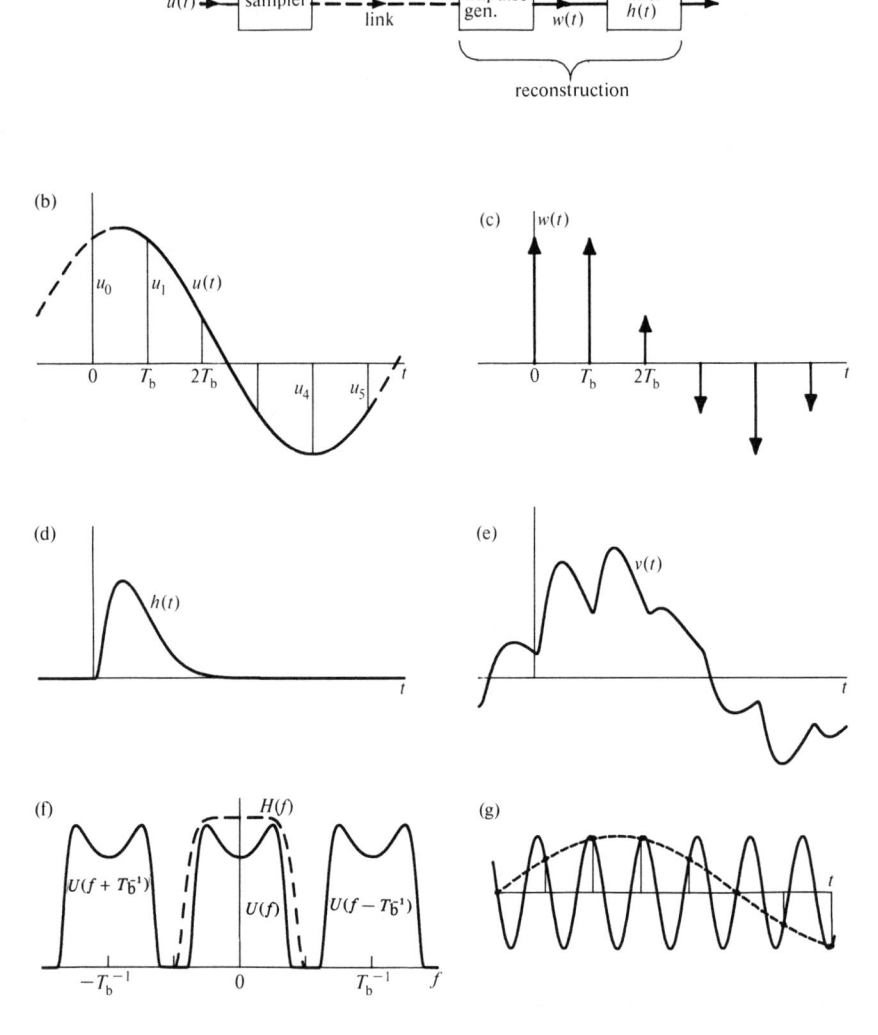

FIG. 3.3 Nyquist sampling: (a) the process of sampling and reconstruction; (b) the sampled values; (c) delta-comb of impulses proportional to sampled values; (d) impulse-response function of filter; (e) output from filter; (f) obtaining faithful reconstruction—$U(f)$ and $H(f)$ are the Fourier transforms of $u(t)$ and $h(t)$ respectively; (g) effect of sampling at too low a rate.

It will now be demonstrated that if $T_b^{-1} > 2f_{\max}$ and if $h(t)$ is suitably chosen (not as in the figure!) then $v(t)$ is equal to the original $u(t)$. This is proved by showing that their Fourier transforms $U(f)$ and $V(f)$ can be made equal. The function $w(t)$ may be written as

$$w(t) = \Sigma_k u(kT_b)\delta(t-kT_b) = u(t)\Sigma_k \delta(t-kT_b) = T_b^{-1} u(t) \text{III}(t/T_b)$$

in terms of the sha function introduced in Section 2.4.2. Hence the Fourier transforms satisfy (Table 2.2)

$$W(f) = T_b^{-1} U(f) * \text{III}(T_b f)$$

with

$$\text{III}(T_b f) = T_b^{-1} \Sigma_n \delta(f - n/T_b).$$

Hence the convolution is

$$W(f) = T_b^{-1} \Sigma_n U(f - n/T_b)$$

and so

$$V(f) = H(f) T_b^{-1} \Sigma_n U(f - n/T_b).$$

A faithful reconstruction of $U(f)$ and hence of $u(t)$ is obtained if

(a) the shifted curves $U(f - T_b^{-1})$, $U(f + T_b^{-1})$ etc. do not overlap significantly with $U(f)$, and (b) the product with $H(f)$ 'blacks out' the shifted versions of $U(f)$. Thus $U(f)$ should be negligible for $|f| \geq \frac{1}{2}T_b^{-1}$ and $H(f)$ should be a low-pass characteristic with a cut-off at $|f| = \frac{1}{2}T_b^{-1}$ (Fig. 3.3(f)).

If we regard $U(f)$ as given rather than T_b, then this means that three conditions must be satisfied:

(a) $U(f)$ should be band-limited to some frequency range $|f| < f_{max}$, by preliminary low-pass filtering if necessary;

(b) T_b should be chosen short enough to give $T_b^{-1} > 2f_{max}$; that is, the sampling rate should exceed $2f_{max}$. (Fig. 3.3(g) shows what can happen if this condition is not satisfied.)

(c) The low-pass filtering $H(f)$ must suppress the repeated versions of $U(f)$. (Fig. 3.3(e) shows what can happen if this condition is not satisfied.)

Thus it is possible to reproduce faithfully a band-limited function by sampling at a fast enough rate.

So far it has been assumed that the sampling of the input signal $u(t)$ is instantaneous. Each sampled value is more likely to be an average of $u(t)$ over the immediately preceding interval. The effect is just the same as if the signal is first passed through a sliding-average filter (averaging over this interval) before instantaneous sampling is used. Thus the overall effect is identical to employing prefiltering, whose effect is not likely to be significant in comparison with the low-pass filtering which is anyway needed to make the sampling work.

3.2.6 *Analogue-to-digital conversion*

Sampling an analogue signal $u(t)$ gives a stream of sampled values $u_k = u(kT_b)$, where T_b is the sampling interval and k takes all integral values. We may then set the height of a pulse in the kth interval to u_k. The possible values form a continuum. This gives us Pulse-Amplitude Modulation (PAM). One reason for doing this is that we may interleave several pulse-trains provided of course that the sampling rate for each train is the same and provided the pulses are made brief enough in time to avoid overlap. This method of sending several signals over the same channel is called time-division multiplexing (TDM). Alternatively once we have the sampled values we may represent them as bit-strings in the same way that numbers are represented as bit-strings on a computer. The total possible range of values of the real signal-variable is broken up into subranges I_l (not necessarily of equal length). The integer l or the corresponding bit-string is output whenever u_k lies in I_l (Fig. 3.4). This process is known as *quantization*. The

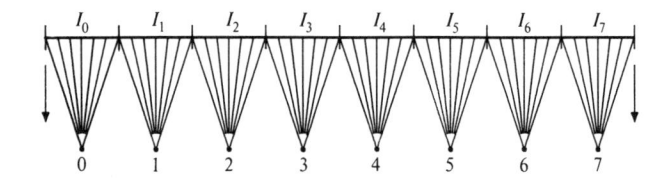

FIG. 3.4 Quantizing a continuous variable: when the variable lies in the range I_l the quantized value is the integer l.

inherent rounding errors are called quantization errors, and these produce noise and other forms of distortion which can be reduced by using more levels, that is, a larger number of bits. This is a typical example of irreversible data-compression as described in Section 3.1.3. Thus on a telephone channel quantization into 256 levels needing 8 bits to represent them is standard, while in high-fidelity digital recording 16 bits per sample is employed. Encoding an analogue signal into bit-strings is known as Pulse-Code Modulation (PCM). The advantage of this form of transmission is that these bit-streams may be coded, encrypted, switched, interleaved, and generally processed by digital electronics just like any other form of digital data. Moreover, such signals may be regenerated at digital repeaters on long-distance links, without noise being amplified at the same time (Section 3.2.7).

Although the voltage intervals in Fig. 3.4 are shown as being equal it is often better to use finer divisions near the zero signal-level and coarser divisions at high signal-levels, in order that 'quiet passages' may be reproduced more faithfully. This process is known as *companding* (Schwartz 1980, Chapter 3).

Example. Telephone signals are filtered so that $f_{max} = 3.5$ kHz. Hence the sampling rate must exceed 7 kHz, i.e. 7000 samples/s. It is usually taken as 8 kHz. With 8-bit PCM this means 64 kilobits/s. For two-channel high-fidelity with $f_{max} = 20$ kHz and 16-bit PCM we would need 1.28 megabits/s.

It should be noted that Fig. 3.4 also illustrates the decoding process for multilevel pulse signalling as described in Section 3.2.4, which assigns a possible message to a range of values of the received signal. This idea can be extended: some irreversible data-compression systems are based on the use of decoders for error-correcting codes (McEliece 1977).

An alternative method of turning analogue signals into digital is known as *delta modulation* (Schwartz 1980, Section 3.7). For further details on quantization see Gersho (1977).

3.2.7 *Links using repeaters*

On a 'long-haul' cable link (i.e. a link extending over a long distance) the attenuation of the cable may be so large that repeaters are needed at regular intervals to boost the signal. We suppose that pulsed signals are being sent. The performance of the link when linear amplifiers are used will be compared with the performance when digital pulse-regenerators are used. This will illustrate the error-correcting power inherent in digital signalling. Let us suppose that the overall link consists of n sections. Each section is a length of cable followed by a booster, a linear amplifier in the first case (Fig. 3.5). The input signal has

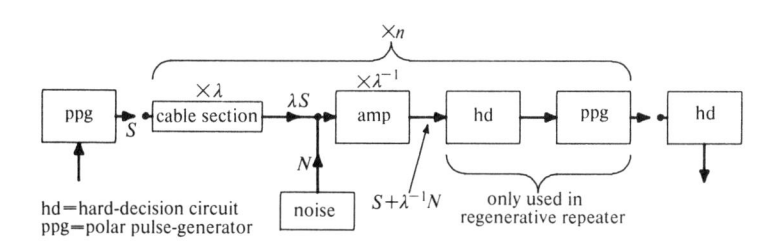

FIG. 3.5 Binary digital link using n repeaters—S is the signal-strength, N the noise-power, and λ is the attenuation of each section of cable.

a power S which is attenuated by the cable down to λS where λ is the attenuation, a positive number very much less than unity. The amplifier has a power gain of $1/\lambda$ which restores the signal back to its original level. Thus the overall gain of each section is unity. Unfortunately the amplifier is noisy and this is represented by the injection at its input of a noise-signal with power N, which is amplified up to $\lambda^{-1}N$ at the output of the section. This noise is then transmitted through each succeeding section just as though it were part of the signal, and finally emerges at the same power-level $\lambda^{-1}N$. However, each section contributes its own noise, so that the total noise-power level at the output is $n\lambda^{-1}N$. (It should be noted that with uncorrelated noise-sources the powers add.)

Let us now imagine that the input is a polar pulse-train, and that at the end of the whole cable there is a *hard-decision* circuit which tries to decide whether each pulse, now corrupted by noise, is '+' or '−'. Let us also assume that this circuit makes errors with a probability $\exp(-S/N_0)$ where S is the output signal power and N_0 is the output noise-power from the cable. (This formula approximates well enough for our purpose what happens with polar signalling in the presence of Gaussian noise. See Section 4.3.) Thus the error-probability for the overall link using linear amplifiers is

$$p_L = \exp\{-\lambda S/(nN)\} = \alpha^{1/n} \qquad (3.5)$$

where

$$\alpha = \exp(-\lambda S/N). \qquad (3.6)$$

Suppose on the other hand that each linear amplifier is followed by a regenerator, which decides on the pulse and then produces a brand-new corresponding pulse. Then there is a chance that at each regenerator a pulse may be passed on erroneously. The error-probability for a single stage is α, as given by (3.6). Then it is readily shown that the overall error-probability, p_R, is less than $n\alpha$ (a close result if $n\alpha$ is small).

We may now compare the two systems. If α is 10^{-10}, then in the first system the error probability is 10^{-1} after 10 links, whereas in the second it is 10^{-9}. The difference arises as follows: in the former case small deviations in the signal are allowed to accumulate and so the chance of obtaining a deviation large enough to deceive the decision circuit is high. In the latter system small deviations are corrected as they occur, and overall we have to be very unlucky if a mistake is made.

3.3 Band-restricted signalling

3.3.1 *Band-restricted pulse signalling*

In most situations the band of frequencies occupied by the signal must be restricted. It is in fact impossible to have a strict bandwidth limitation for a pulse of finite duration, but any slight overspill may usually be regarded as adding to the general noise-level for users of adjacent frequencies. Conversely if a receiver is not strictly band-limited then its reception is interfered with by perfectly legitimate users of adjacent frequencies, but this again can usually be lumped in with the inevitable noise.

A typical signal consisting of a train of pulses sent at a rate T_b^{-1} may be written

$$s(t) = \Sigma_k b_k y(t - kT_b) \qquad (3.7)$$

where $y(t - kT_b)$ is a pulse function $y(t)$ shifted to the slot k. Thus for instance $y(t)$ might be a rectangular pulse (2.3)

BAND-RESTRICTED SIGNALLING

$$\Pi(t/T_b) = \begin{cases} 1 & \text{for } |t| < \tfrac{1}{2}T_b, \\ 0 & \text{for } |t| \geqslant \tfrac{1}{2}T_b. \end{cases}$$

The information in the signal is carried by the values of the coefficients b_k. The range of values available to b_k may be 0 and 1 for on–off keying, or a continuous range for pulse-amplitude modulation. At the moment what this range is is not significant. The choice $\Pi(t/T_b)$ for $y(t)$ satisfies

$$y(kT_b) = \begin{cases} 1 & \text{for } k = 0, \\ 0 & \text{for any non-zero integral value of } k. \end{cases} \quad (3.8)$$

Hence the values of b_k are immediately determined from $s(t)$ by

$$s(lT_b) = \Sigma_k b_k y\{(l-k)T_b\} = b_l, \quad (3.9)$$

since $y\{(l-k)T_b\}$ vanishes unless $k = l$, when it is 1. The discontinuities of rectangular pulses give rise to a $1/f$ behaviour in the frequency spectrum at large $|f|$, and thus such pulses are unsuitable if there are restrictions on the bandwidth. There are other functions $y(t)$ which obey the condition (3.8) and hence permit the use of sampling to determine the signal coefficients.

(a) The effect of rolling off the frequency spectrum by multiplying by say the bell-shaped 'raised-cosine' function $R(fT_s)$ {given by (2.6)} with a roll-off frequency T_s^{-1} is to convolute the rectangular pulse $\Pi(t/T_b)$ by the inverse Fourier transform of $R(fT_s)$ which has a roll-off at T_s (Fig. 3.6). (See Exercise 3.10.) This rounds the shoulders of the pulse so that it falls off over a time-scale T_s (Fig. 3.6(b)). Provided that this time is very much less than the pulse-width T_b such a pulse will be called a *quasi-rectangular* pulse. None the less the bandwidth of a quasi-rectangular pulse must usually be regarded as

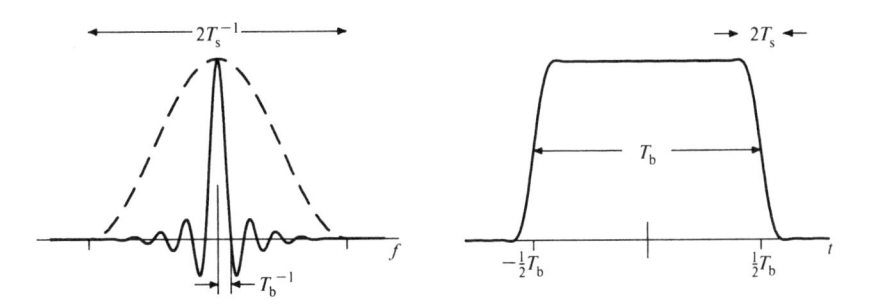

FIG. 3.6 (a) sinc-function spectrum $\text{sinc}(fT_b)$ of a rectangular pulse after the spectrum has been multiplied by the raised-cosine transfer function $R(fT_s)$ (shown dashed); (b) the ensuing quasi-rectangular pulse.

54   DIGITAL SIGNALLING METHODS

excessive, being of order T_s^{-1} rather than T_b^{-1}. (Such a function does not obey (3.8) strictly, but the error is negligible if $T_s \ll T_b$.)

(b) A commonly used pulse shape for $y(t)$ is the raised-cosine function itself, as $R(t/T_b)$ (Equation 2.6), which is non-zero over an interval $2T_b$, twice the pulse-spacing. Since this function is continuous with a continuous first derivative but with a discontinuous second derivative its Fourier transform falls off as f^{-3} for large $|f|$. Moreover it obeys the condition (3.8).

(c) The sinc-pulse $\text{sinc}(t/T_b)$ provides an example of a strictly band-limited signal since

$$\text{sinc}(t/T_b) = T_b \int_{-1/(2T_b)}^{1/(2T_b)} \exp(2\pi j f t) \, df. \quad (3.10)$$

This function also obeys (3.8). Of course it suffers from a very slow fall-off in t, and so one would normally multiply it by some roll-off function such as the raised cosine $R(t/T_c)$ with $T_c \gg T_b$ (Fig. 3.7(a)), without its losing the property (3.8). (The effect of this is to broaden the spectrum beyond $\pm\frac{1}{2}T_b^{-1}$ by about T_c^{-1} (Fig. 3.7(b)).)

(d) We may even multiply the original sinc-function by itself to give double zeros, and a triangular frequency spectrum which stretches from $-T_b^{-1}$ to $+T_b^{-1}$. The double-zeros reduce the likelihood of *intersymbol interference* (ISI) caused by poorly timed sampling.

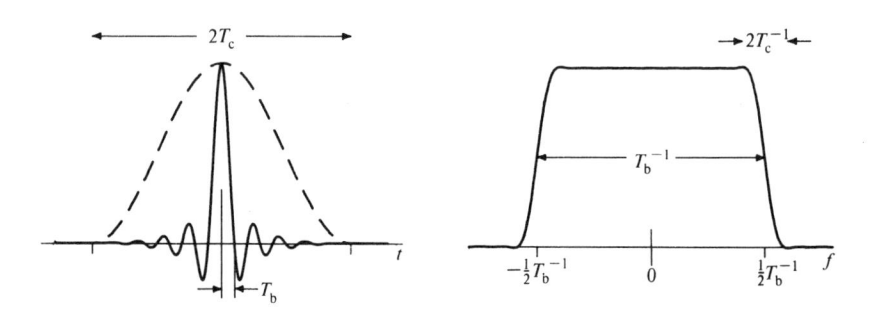

FIG. 3.7 (a) sinc-like pulse function $\text{sinc}(t/T_b) R(t/T_c)$; (b) the corresponding frequency-spectrum. Note that these curves are duals of those in Fig. 3.6.

The frequency spectrum $Y(f)$ of any function $y(t)$ satisfying (3.8) possesses the Nyquist property

$$\Sigma_n Y(f - nT_b^{-1}) = 1. \quad (3.11)$$

This is because

$$y(t) \text{ III}(t/T_b) = y(t) T_b \Sigma_k \delta(t - kT_b) = T_b \Sigma_k y(kT_b) \delta(t - kT_b) = T_b \delta(t).$$

Thus we find by taking the Fourier transform

$$Y(f) * T_b \text{III}(T_b f) = T_b \cdot 1$$

and hence we obtain (3.11).

The attentive reader may well wonder why instantaneous sampling (3.9) has been advocated for the band-limited functions whereas averaged sampling was recommended for rectangular pulses, on the grounds that instantaneous sampling was too susceptible to noise. The process of 'integrate-and-dump' is however equivalent to filtering (by a filter with a rectangular impulse response function) followed by instantaneous sampling (see Section 4.3.2). Thus it is certainly legitimate to use instantaneous sampling even for rectangular pulses, provided that the sampler is preceded by the appropriate filter. For band-limited pulses instantaneous sampling can again be used at the receiver, provided that the sampler is preceded by a band-limiting filter that removes out-of-band noise, without causing the frequency spectrum of each pulse to violate the Nyquist criterion (3.11).

Finally we compare the bandwidths required to send a bandlimited signal (a) directly, (b) by PAM, and (c) by binary PCM. Direct transmission requires a bandwidth of f_M, the band-limiting frequency. In order to send the signal by PAM we must sample at a rate T_b^{-1} greater than $2f_M$, and if the pulse waveforms $y(t)$ in (3.7) are the sinc functions given in (3.10), then the bandwidth is $\frac{1}{2}T_b^{-1}$ which is greater than f_M. In practice $\frac{1}{2}T_b^{-1}$ will exceed f_M by 20 per cent at least and the pulse wave-forms will not be ideal sinc-functions for the reasons discussed above. Thus the bandwidth will exceed $\frac{1}{2}T_b^{-1}$ by another 20 per cent, say (Fig. 3.7(b)). Thus we may expect that PAM transmission will need something like a 50 per cent increase in bandwidth above the original value f_M.

If the signal is sent by PCM using an n-bit binary code then the bandwidth is further expanded by a factor n. For telephone signals and for digital TV n is typically chosen to be 8, and larger values up to 16 are used for high-fidelity sound. Thus we see that PCM requires an expansion of the bandwidth by well over an order of magnitude. 16-bit PCM transmission of high-fidelity stereo sound using two channels of 20 kHz each requires a bandwidth of the order of 2 MHz (Carasso, Peek, and Sinjou 1982). PCM transmission of television with a bandwidth of about 6 MHz requires a bandwidth not far short of 100 MHz. Thus the techniques developed for handling analogue television signals have ample bandwidth for PCM coding of high-fidelity sound, and we may expect that PCM will become the standard (Section 1.1.3). On the other hand digital transmission of broadcast-quality television is not so likely in the immediate future, since it requires a wide bandwidth and fast digital electronics. (But see Stafford 1980, Tomlinson 1983.)

3.3.2 Intersymbol interference

Intersymbol interference (ISI) can arise as a consequence (a) of bad timing and (b) of filtering, whether this is deliberately put into the receiver circuits, or whether it is caused by dispersion in the channel. In the latter case it may vary with time. Suppose that a pulse $y(t)$ satisfying (3.8) is convoluted with some filter impulse-response function $h(t)$. A new function $y' = y * h$ is obtained which in general no longer obeys (3.8). Thus the pulse-train (3.7) is replaced by $s' = s * h$. Hence we find

$$s'(t) = \Sigma_k b_k y'(t - kT_b)$$

and the sampled values are

$$d_l = s'(lT_b) = \Sigma_k a_{l-k} b_k = \Sigma_k a_k b_{l-k}$$

with $a_{l-k} = y'\{(l-k)T_b\}$. For simplicity it will be assumed that $a_0 = 1$, so that

$$d_l = b_l + a_1 b_{l-1} + a_2 b_{l-2} + \ldots$$

where hopefully the coefficients $a_1, a_2 \ldots$ are small and well known. If time-division multiplexing is being used this interference in one pulse from its predecessors will lead to 'cross-talk'. The correction may be performed recursively by $b_l = d_l - a_1 b_{l-1} - a_2 b_{l-2} - \ldots$ so that the corrected value b_l is obtained from the received value and the previously corrected values $b_{l-1}, b_{l-2}, \ldots$ fed back in again. If the a_i are not small this method is liable to show unstable behaviour like any system with too much feedback.

Circuits which correct for ISI are called *equalizers*. Those that attempt to compensate for the effects of a slowly varying channel are called adaptive equalizers (Section 3.3.3).

3.3.3 Adaptive equalizers

It is interesting to see how the error-correcting capabilities of digital signalling may be used to follow and compensate for the characteristics of a slowly varying channel. A filter which does this is known as an *adaptive equalizer*.

One type uses a *transversal filter*, a chain of delays each of duration T_b (Fig. 3.8). The tapped outputs are then multiplied by 'constants' c_i and summed. Thus with n delays the output v_k is related to the input u_k by

$$v_k = \Sigma_{l=0}^{n} c_l u_{k-l}.$$

The problem is how to keep changing the settings of the c_k to follow slow changes in the channel.

The error (an analogue value by the way!) is $e_k = v_k - s_k$, where s_k is the true signal-sequence at the transmitter. Now provided the equalizer is reasonably well set up, the decision circuit at the output will almost certainly provide the correct output sequence s_k, but we cannot guarantee that this happy state of

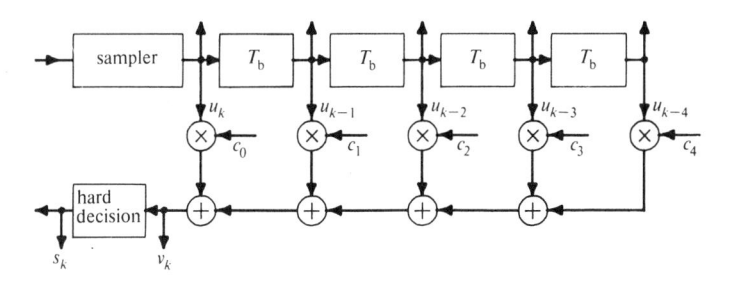

FIG. 3.8 Transversal filter forming part of adaptive equalizer.

affairs will be maintained if the c_k are not suitably adjusted as the characteristics of the channel change.

For a moment we consider another problem: suppose we wish to minimize some differentiable function $F(\mathbf{r})$ in three dimensions. One way is to start at some point $\mathbf{r}_0$ and move in the direction opposite to the gradient to a new point $\mathbf{r}_1 = \mathbf{r}_0 - \Delta \nabla F$ where the gradient is evaluated at $\mathbf{r} = \mathbf{r}_0$. The quantity Δ controls the step-length. (In other words the ith coordinate of $\mathbf{r}_0$ is incremented by $-\Delta \cdot \partial F/\partial x_i$.) Then the process is iterated, starting anew at $\mathbf{r}_1$. If Δ is too small the process converges very slowly. If it is too large we may overshoot the minimum and the sequence of points $\mathbf{r}_i$ is liable to oscillate or become unstable.

In the case of the adaptive equalizer we use this method at each step to reduce the error, or rather the square of the error, which is mathematically easier to handle. Thus at the kth step each coefficient c_m is changed by $-\Delta \partial(e_k^2)/\partial c_m$, where as before Δ controls the step-length. This change is given by

$$-\Delta \cdot \partial(e_k^2)/\partial c_m = -2\Delta \cdot e_k \partial e_k/\partial c_m.$$

Since

$$e_k = (\Sigma_{l=0}^{n} c_l u_{k-l}) - s_k,$$

the derivative is $\partial e_k/\partial c_m = u_{k-m}$, because u_{k-m} is the coefficient of c_m. Thus at the kth step each coefficient c_m is changed by $-2\Delta \cdot (v_k - s_k)u_{k-m}$. It is evident from Fig. 3.8 that all the quantities needed to evaluate this expression are available to the filter-control. If Δ is too small the equalizer may not be able to follow changes in the channel sufficiently rapidly. If it is too large the equalizer may overstep the minimum-error point and become unstable. It is crucial for the success of this method that the correct sequence s_k be available with high reliability. Provided Δ is not too large the occasional incorrect decision about s_k will not cause the equalizer to get lost.

At the beginning of the transmission the adaptive equalizer has to be 'trained' to match the channel. A sequence known to the receiver is sent and thus the quantities s_k are available, even if the equalizer is so badly adjusted that the decision-circuit outputs are useless. While the training sequence lasts the coefficients can be adjusted exactly as described above.

Adaptive equalizers are a very important component in digital signalling and we have only touched the theory. Further details may be found in Lucky, Salz, and Weldon (1968), Proakis (1983), and in Qureshi (1982).

3.3.4 Duobinary system

To improve the frequency spectrum of a pulse signalling system we may introduce deliberate intersymbol interference which is removed later. (Unfortunately this does not necessarily make the system any less susceptible to extra intersymbol interference caused, say, by dispersion in the channel.) Thus suppose a modulator forms the signal (3.7)

$$s(t) = \Sigma_k b_k y(t - kT_b) \qquad (3.12)$$

from an input stream of real values b_k. Here $y(t)$ is a suitable pulse-shape. Now suppose that instead of the b_k we feed in the values

$$d_k = b_k + b_{k-1} \qquad (3.13)$$

produced by an 'encoder', to give the modified signal

$$s'(t) = \Sigma_k d_k y(t - kT_b). \qquad (3.14)$$

(We start at the beginning, $k = 0$ say, with $b_{-1} = 0$.) Then it is readily seen that

$$s'(t) = s(t) + s(t - T_b) \qquad (3.15)$$

so that the system is equivalent to the original system followed by a filter with impulse-response function

$$h(t) = \delta(t) + \delta(t - T_b).$$

The transfer function of this hypothetical filter is

$$H(f) = 1 + \exp(2\pi j f T_b)$$

with a magnitude $2|\cos(\pi f T_b)|$ which vanishes at $f = \pm \frac{1}{2} T_b^{-1}$. As explained above the original frequency spectrum usually has a cut-off around $\frac{1}{2} T_b^{-1}$ and so this filtering reduces still further the frequency components in the vicinity of the cut-off. This effectively reduces the bandwidth and it also reduces the errors caused by poorly timed sampling at the receiver. To recover the original values b_k at the receiver we first sample $s'(t)$ to determine the d_k, and then 'decode' by using

$$b_k = d_k - b_{k-1}.$$

BAND-RESTRICTED SIGNALLING

A serious disadvantage of this system is that an error in receiving one of the b_k is propagated on to its successors. In the case we now consider, where the b_k take only two values 0 and 1, this situation will eventually be detected by a computed value of a b_k being out of its possible set of values. Thus the two b_k sequences shown below differ in every place, and yet the corresponding encoder outputs d_k differ in only one place.

$$0\,|\,1\,0\,1\,0\,1\,0\,1\,0\,1\,0\,b_k \qquad\qquad 0\,|\,0\,1\,0\,1\,0\,1\,0\,1\,0\,1\,b_k$$
$$|\,1\,1\,1\,1\,1\,1\,1\,1\,1\,1\,d_k = b_k + b_{k-1} \qquad |\,0\,1\,1\,1\,1\,1\,1\,1\,1\,1\,d_k.$$

A way round this problem is to generate the signal (3.14) with the d_k encoded according to

$$c_k = b_k \oplus c_{k-1}, \qquad d_k = c_k + c_{k-1}, \qquad (3.16)$$

instead of (3.13). (The initial conditions are $c_{-1} = 0$. The symbol $\oplus$ denotes binary addition without carry, so that $1 \oplus 1 = 0$.) The extra pre-processing replaces the input string b_k to the previous system by c_k, and so does not affect the suppression of frequency components in the vicinity of $\frac{1}{2}T_b^{-1}$. Now the decoding is very simple: if $d_k = 1$ then $b_k = 1$, and if $d_k = 0$ or 2 then $b_k = 0$ (Exercise 3.13). Thus there is no possibility of error propagation, and yet the suppression of high-frequency components still occurs. The operation of the encoder may be illustrated in a different way. It may be in one of two states S_0 and S_1 as shown in Fig. 3.9. If an input $b_k = 0$ comes in, then it follows the solid line out of its state back to the same state. At the same time it puts out the coding $d_k = 0$ or 2 as indicated. But if an input $b_k = 1$ comes in, then it follows the dotted line to the other state. For either transition it puts out the coding 1.

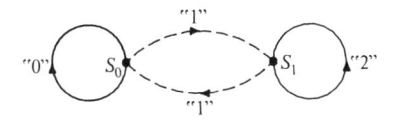

FIG. 3.9 Encoder for the duobinary code. Solid lines denote transitions caused by input 0, dashed lines by input 1. The symbols '0', '1', and '2' are the outputs.

This encoding gives rise to some redundancy in the output since each symbol can in principle convey $\log_2 3 = 1.585$ bits. Associated with this there is some error-correcting potential which is not used by the decoding algorithm just described (see Section 6.2.1).

For further details on this and related forms of signalling see Pasupathy (1977).

3.4 Carrier-based signalling

3.4.1 *Carrier modulation*

For many purposes it is convenient to use the basic signal $a(t)$ to 'modulate' a carrier proportional to $\cos(2\pi f_c t)$ where the carrier-frequency f_c is greater than the frequencies of the Fourier components making up $a(t)$. It will be assumed that these frequencies are restricted to some range $-f_M$ to $+f_M$, with the maximum modulation frequency f_M less than f_c. In other words the Fourier transform $A(f)$ of $a(t)$ is negligible for $|f| > f_M$ and $a(t)$ is said to be band-limited to frequencies below f_M. Typically for band-restricted pulse signalling we would expect $f_M \approx T_b^{-1}$. Conceptually the simplest form of modulation (known as *amplitude modulation* or AM) is to multiply $a(t)$ by $2\cos(2\pi f_c t)$ to give the real signal

$$s(t) = a(t)\exp(2\pi j f_c t) + a(t)\exp(-2\pi j f_c t).$$

The original signal $a(t)$ is called the baseband signal, while $s(t)$ is the carrier-based signal. The Fourier transform of $a(t)\exp(2\pi j f_c t)$ is $A(f - f_c)$ by Table 2.2, and hence the Fourier transform of $s(t)$ is $A(f - f_c) + A(f + f_c)$. Since $A(f)$ is negligible for $|f| > f_M$, the frequency components of $s(t)$ on the positive-frequency side must lie between $f_c - f_M$ and $f_c + f_M$. There are two reasons for modulating a carrier. First it is much easier to send high-frequency signals by radio, and secondly the same medium can carry several signals distinguished by their carrier frequencies. Provided the frequency bands do not overlap the signals may be separated out by band-pass filters. This technique of using several carrier-frequencies, one for each signal, on a common channel is called Frequency-Division Multiplexing (FDM). At the receiving end the carrier-based signal is multiplied by $\cos(2\pi f_c t)$, to give $s'(t) = a(t) + a(t)\cos(4\pi f_c t)$. The second term consists only of components with frequencies around $\pm 2f_c$, and so can be removed by a low-pass filter. It should be noted that if the second multiplication is carried out with $\sin(2\pi f_c t)$, then the output from the low-pass filter is zero (Exercise 3.15). Thus in this system of demodulation the phase of the local oscillator providing the multiplier-function must be carefully controlled so that it is synchronized with the oscillator in the transmitter. The ordinary large-carrier system of amplitude modulation is similar, but a constant a_0 chosen so that $a(t) + a_0$ is always positive is added to $a(t)$ before modulation so that at the receiver the simply implemented technique of 'envelope' demodulation may be used. Here the receiver in effect regenerates $a(t)$ by putting a smooth curve through the peak values of the modulated carrier. In this system there is no need for synchronization.

In most applications the maximum modulation frequency f_M is very much less than f_c. Such a system is called 'narrow-band' (Section 1.2.1). We shall usually assume that this is the case. There are however some systems which do not have this property:

CARRIER-BASED SIGNALLING

(a) VLF communications with submarines require the use of very low carrier-frequencies (around 10 kHz and even lower). Hence demanding $f_M \ll f_c$ would give very low signalling rates.

(b) Spread-spectrum systems may use a value of f_M comparable with f_c (Section 5.3.6). See also Harmuth (1981).

(c) Voice-grade telephone circuits only pass frequencies in the range 0.3 to 3.5 kHz. Thus for sending carrier-based signals by telephone it is convenient to choose a carrier-frequency near the middle of this band, at 1.9 kHz say, and modulate it at frequencies up to $f_M = 1.6$ kHz.

3.4.2 Positive-frequency and phasor representations

In most cases the maximum modulation frequency f_M is less than the carrier-frequency f_c. Then it is very convenient to break up the spectrum $S(f)$ of such a signal $s(t)$ as follows:

$$s(t) = \int_{-\infty}^{\infty} S(f) \exp(2\pi j f t) \, df$$

$$= \int_{\epsilon}^{\infty} S(f) \exp(2\pi j f t) \, df + \int_{-\infty}^{-\epsilon} S(f) \exp(2\pi j f t) \, df$$

$$+ \int_{-\epsilon}^{\epsilon} S(f) \exp(2\pi j f t) \, df.$$

Here ϵ is less than $f_c - f_M$ so that the third term which contains the low-frequency part of the spectrum can be completely neglected. The second term on the right-hand side is the complex conjugate of the first (Exercise 3.16), and so we may set

$$s(t) = 2^{-1/2} z(t) + 2^{-1/2} z^*(t) \qquad (3.17)$$

with

$$z(t) = \sqrt{2} \int_{\epsilon}^{\infty} S(f) \exp(2\pi j f t) \, df. \qquad (3.18)$$

(The reason for introducing the factor $\sqrt{2}$ will become apparent in a moment.) Thus the Fourier transform $Z(f)$ of $z(t)$ vanishes if $f < \epsilon$ and is equal to $\sqrt{2}.S(f)$ if $f \geqslant \epsilon$. (A similar decomposition into positive- and negative-frequency parts can be made of the Fourier series of a real periodic function provided the zero-frequency component is zero.) The representation (3.17) of a signal $s(t)$ by means of a function $z(t)$ satisfying (3.18) will be called the *positive-frequency representation*. It is very useful when there is a naturally occurring gap between the positive and negative frequency parts of the spectrum $s(t)$. Conversely it is not useful when there is no such gap, as in base-band signalling, or when the maximum modulation frequency is comparable with the carrier frequency. Under these circumstances we may still use the base-band Fourier methods previously described, which are of course completely general.

Since the frequency components of $z(t)$ are close to a carrier-frequency f_c, we see that $z(t)$ is a rapidly rotating vector in the complex plane rotating at around f_c revolutions/s in an anticlockwise sense. A function which varies in a much more leisurely manner is the *phasor*

$$\phi(t) = \exp(-2\pi j f_c t) z(t). \tag{3.19}$$

Its Fourier transform satisfies

$$\Phi(f) = \begin{cases} \sqrt{2}.S(f - f_c) & \text{for } f > \epsilon - f_c, \\ 0 & \text{for } f \leqslant \epsilon - f_c. \end{cases} \tag{3.20}$$

This Fourier transform is negligible outside the range $-f_M$ to $+f_M$, but since $\phi(t)$ is not in general real, the transform $\Phi(f)$ need not satisfy $\Phi(-f) = \Phi^*(f)$. Thus we find

$$s(t) = 2^{-1/2} \exp(2\pi j f_c t) \phi(t) + 2^{-1/2} \exp(-2\pi j f_c t) \phi^*(t), \tag{3.21}$$

and writing $\phi(t) = \rho(t) \exp\{j\theta(t)\}$ we obtain

$$s(t) = \sqrt{2}.\rho(t) \cos\{2\pi f_c t + \theta(t)\}.$$

Thus the magnitude of $s(t)$ is proportional to the magnitude of $\phi(t)$ and the phase is given by the phase of $\phi(t)$.

It is found that the total energy satisfies

$$\int_{-\infty}^{\infty} \{s(t)\}^2 \, dt = \int_{-\infty}^{\infty} z^*z \, dt = \int_{-\infty}^{\infty} \phi^*\phi \, dt. \tag{3.22}$$

In terms of z the left-hand side is

$$\tfrac{1}{2} \int_{-\infty}^{\infty} z^2 \, dt + \tfrac{1}{2} \int_{-\infty}^{\infty} z^{*2} \, dt + \int_{-\infty}^{\infty} z^*z \, dt$$

and the first two integrals vanish. This is because

$$\int_{-\infty}^{\infty} z^2 \, dt = 2 \int_{\epsilon}^{\infty} df' \int_{\epsilon}^{\infty} df'' S(f') S(f'') \int_{-\infty}^{\infty} \exp[2\pi j(f' + f'')t] \, dt,$$

and the time-integral is $\delta(f' + f'')$, which must be zero since $f' + f'' \geqslant 2\epsilon$. The second integral vanishes for the same reason. The second equality in (3.22) follows immediately from (3.19). The reason for introducing the factor $\sqrt{2}$ in the definition of $z(t)$ and of $\phi(t)$ is to avoid factors of 2 appearing in the expression for the energy in terms of $|z|^2$ or of $|\phi|^2$.

3.4.3 Quadrature modulation

The phasor $\phi(t)$ may be used to represent two real base-band signals $a(t)$ and $b(t)$ by

$$\phi(t) = a(t) + jb(t). \tag{3.23}$$

In this way we may put two signals on the same carrier. The r.f. signal $s(t)$ is given by (3.21) and (3.23). At the receiver we recover $\phi(t)$ by multiplying $s(t)$ by $\sqrt{2}.\exp(-2\pi jf_c t)$ to give $\phi(t) + \phi(t)\exp(-4\pi jf_c t)$. The second term is rejected by a low-pass filter with a bandwidth from $-f_M$ to $+f_M$. Since filters do not process complex signals this process is implemented by two identical filters as shown in Fig. 3.10. If the local oscillator is not properly synchronized then the outputs are 'matrixed'. It should be emphasized that this syncrhonization involves timing accuracies of the order f_c^{-1}, whereas the avoidance of intersymbol interference in the base-band signal requires timing accuracies of the order of T_b or of f_M^{-1}, which are usually easier to achieve.

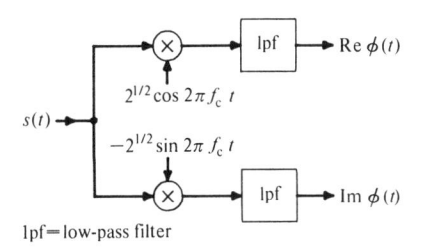

FIG. 3.10 Synchronous demodulation of a carrier-based signal to obtain real and imaginary parts of the phasor.

3.4.4 Carrier-based pulse signalling

In the phasor representation we may use (3.21) with

$$\phi(t) = \Sigma_k c_k u(t - kT_b). \tag{3.24}$$

Here c_k and $u(t)$ may be complex. (In order for the phasor representation to be valid, the Fourier transform $U(f)$ of the function $u(t)$ must obey a restriction like (3.20), that $U(f) = 0$ for $f < \epsilon - f_c$ with ϵ some positive quantity. This means in practice that pulses must be switched on or off over a time-scale of several cycles of the carrier. The restriction that the bandwidth of u be much less than the carrier-frequency is much stronger and quite sufficient.) We may use the real and imaginary parts of c_k to convey separate information streams. Careful synchronization has to be maintained since timing errors of the order of f_c^{-1} can cause cross-talk between these streams.

Various ways in which the allowable values of c_k are distributed around the complex plane are shown in Fig. 3.11. Such patterns are known as *signal constellations* (Schwartz 1980, Section 4.3). In On–Off Keying (OOK) there are only two values, one being zero (Fig. 3.11(a)). In binary phase-shift keying (BPSK), also known as phase-reversal keying (PRK), there are two values of

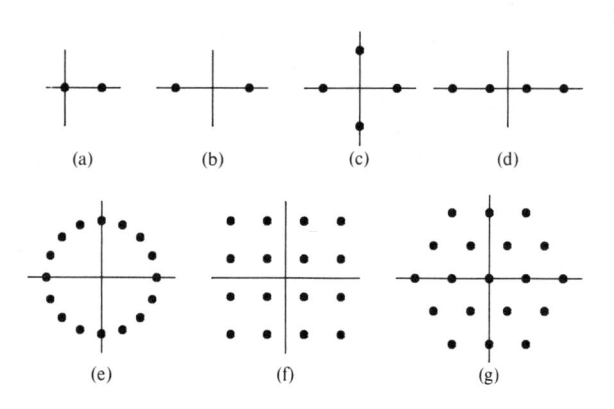

FIG. 3.11 Constellations for carrier-based pulse-signalling: (a) OOK; (b) BPSK or PRK; (c) QPSK; (d) 4-ASK; (e) 16-PSK; (f) QAM; (g) hexagonal.

opposite phase (Fig. 3.11(b)). This system is analogous to the polar system for base-band signalling, but whereas there is no difficulty in distinguishing positive and negative polarities, a loss of synchronization in the r.f. system can cause the interchange of 0 and 1. In quadrature phase-shift keying (QPSK) there are four values (Fig. 3.11(c)), as there are in four-level amplitude-shift keying (ASK) (Fig. 3.11(d)). Obvious generalizations of these systems are multiple phase-shift keying (MPSK), where there is some number m of points arranged evenly around a circle centred on the origin (Fig. 3.11(e)), and m-level ASK where there are m points along a straight line through the origin. The permitted amplitudes may form two-dimensional arrays as shown in Fig. 3.11(f) (quadrature amplitude shift keying (QASK) or quadrature amplitude modulation (QAM)). A hexagonal constellation of 19 points is shown in Fig. 3.11(g). This system has the disadvantage that the number of points is not a power of 2, but it is marginally better than the system of Fig. 3.11(f) in that the points are denser for the same minimum distance, so that the system carries a little more information for the same susceptibility to noise.

Determining the actual values of c_k from $s(t)$ is straightforward in principle. If the $u(t)$ in (3.24) satisfy the sampling condition (3.8), then we may find the real and imaginary parts of each c_k by sampling the outputs of the circuit in Fig. 3.10 at the times $t = kT_b$. Thereafter it is a matter of determining the point of the constellation nearest to $(\text{Re } c_k, \text{Im } c_k)$.

All these methods (except OOK) need accurate carrier-phase tracking. (For details see Spilker 1977, Gagliardi 1978, or Bhargava, Haccoun, Matyas, and Nuspl 1982.) Not so critical is the system of differential phase-shift keying (DPSK) where the phase is shifted by 180° if a 1 is to be sent and is left unchanged if 0 is to be sent. This is very similar to the system of Fig. 3.11(b), but one may obtain a phase-reference for each pulse from its predecessor.

We may estimate the information-rate for moderate reliability as we did for base-band signalling when the signal-to-noise ratio is large. In signal constellations like those of Fig. 3.11(f) and Fig. 3.11(g) we would choose the point spacing to be of the order of $\alpha\sqrt{N}$, where N is the noise-power, and the overall radius as proportional to $\alpha\sqrt{S}$, where S is the average signal power. Here α is a constant of proportionality whose value is not of any concern here. Thus the number of points in the constellation is of order S/N and with T_b^{-1} pulses/s the information rate is roughly $T_b^{-1} \log_2 (S/N)$, which is twice what we obtained for base-band signalling. But it should be noted that the bandwidth in the positive frequencies from $f_c - f_M$ to $f_c + f_M$ with $f_M \approx T_b^{-1}$ is twice what was needed in base-band signalling, where it goes from 0 to f_M. As we might expect, the factor $\frac{1}{2}$ reappears for multi-level ASK or for MPSK.

3.4.5 Frequency-shift keying

The travelling-wave tubes that are commonly used for the output stages of microwave transmitters run best at constant power and do not take kindly to amplitude modulation. Hence signalling systems are preferred where the phasor has a constant magnitude. One possible system is to use pulse-modulation (3.24) with quasi-rectangular pulses for $u(t)$ and with the c_k taking values of fixed magnitude but different phases (Fig. 3.11(e)). This method is rather extravagant on bandwidth since the signal itself will contain quasi-discontinuities. On the other hand if band-limited pulse shapes are used for $u(t)$ as discussed in Section 3.3.1, then there is no guarantee that the magnitude of $s(t)$ as given by (3.21) should remain constant (Exercise (3.17).

Instead a form of signalling may be used with the signal given as usual by (3.21) with the phasor moving on a circle maintaining constant magnitude. This is known as *angle modulation*. A simple system of this type is binary frequency-shift keying (BFSK) where the phasor rotates at a constant angular rate during each signalling interval T_b, this rate being $+f_0$ revolutions/s for a 0 and $-f_0$ revolutions/s for a 1. Evidently the angle of the phasor changes by $\pm 2\pi f_0 T_b$ radians in each interval T_b. Thus $\phi(t)$ is a continuous function of t, though with a discontinuous first derivative. The signal too has this property and so its spectrum falls off as f^{-2} for large $|f|$, and its power-spectrum as f^{-4}.

The system is known as frequency-shift keying because in effect the transmitted frequencies are $f_c \pm f_0$ during each interval T_b and hence we use one frequency $f_c + f_0$ to denote 0 and another, $f_c - f_0$, to denote 1.

The smallest convenient use for f_0 turns out to be $f_0 = 1/(4T_b)$, where the phasor rotates through a right-angle one way or the other in each interval T_b. Under these circumstances the system is called minimum-shift keying (MSK) (Pasupathy 1979).

The sampling method of Section 3.3.1 is not of any use for such signals. If the frequencies $f_c \pm f_0$ are sufficiently widely separated, which means in practice that $f_0 T_b \gg 1$, then a pair of narrow-band filters with one filter tuned to each frequency will distinguish the two transmissions. Other methods have to be used for MSK, such as are described in Section 6.2.

For further reading: there are many good texts on communications theory, most of them also covering analogue communications. Here is a selection: Coates (1983), Feher (1981), Gagliardi (1978), Haykin (1978), Lathi (1983), Miller (1983), Peebles (1976), Roddy and Coolen (1981), Roden (1979, 1982), Schrader (1980), Schwartz (1980), Shanmugam (1979), Stanley (1982), Stark and Tuteur (1979), Stremler (1982), Ziemer and Tranter (1976). For tutorial articles see Sklar (1983*a*, *b*). Texts on satellite communications such as Bhargava et al. (1981), Feher (1983), and Spilker (1977) also contain a lot of interesting information.

Exercises

3.1. With $q = 26$ find the value of n which makes q^n equal to 10^{80}, which is a number of the order of the number of atoms in the universe, on current theories.

3.2. Show that if $f(xy) = f(x) + f(y)$ for all $x, y > 0$ and if f is a smooth function then $f(x) = A \ln(x)$ for some A. (Hint: differentiate with respect to y and then set $y = 1$. Finally integrate with respect to x.)

3.3. Estimate the amount of information (in bits) (a) in the *Encyclopaedia Brittanica*, (b) on a 2400 foot (732 m) magnetic tape with 1600 8-bit words per inch (630 words/cm), (c) on a digital TV picture with 800 × 600 picture elements, each requiring 8 bits, (d) on a videotape that can store a 2-hour TV program at 25 pictures (frames) per second, (e) in human DNA, with a 4-symbol alphabet, in which the symbols are spaced at 3×10^{-10} m along a linear molecule of length 1.7 m.

3.4. Show that

$$n! = \int_0^\infty x^n e^{-x} \, dx = \int_0^\infty \exp(n \ln x - x) \, dx.$$

For what value of x is the exponent $n \ln x - x$ a maximum? Expand this

EXERCISES

exponent about the maximum as a Taylor series to second order, and hence approximate the integral by a Gaussian integral, in an approximation good for large n. Hence or otherwise derive Stirling's formula

$$\ln(n!) \approx n \ln n - n + \ln\{\sqrt{(2\pi n)}\} \text{ for large } n.$$

3.5. Show that the number of strings of length N containing the ith character of an alphabet of q characters n_i times is $N!/(n_1! n_2! \ldots n_q!)$. (Here $N = \Sigma_i n_i$.) By assuming that $n_i = p_i N$ with p_i fixed, and by letting N become very large, show that the average information per character is

$$-\Sigma_i p_i \log_2 p_i.$$

3.6. By using induction show that

$$\Sigma_{j=-(n-1)/2}^{j=(n-1)/2} j^2 = n(n^2-1)/12.$$

(Hint: assume true for $n = k$ and prove it for $n = k + 2$. Prove it for $n = 1$ and 2.)

3.7. A binary number is chosen at random in the range 0 to $2^n - 2$ and is then increased by 1. Show that if n is large roughly 2 bits are changed on average.

3.8. The signal $\cos(2\pi f_0 t)$ is sampled at a rate T^{-1}. What output is obtained if the output from the digital-to-analogue converter is put through a low-pass filter with a cut-off at $T^{-1}/2$, in the event that $2f_0 T$ is slightly greater than 1?

3.9. Sketch the pulse-trains corresponding to Fig. 3.1(b) and Fig. 3.1(c) if the raised-cosine pulse-shape described in Section 3.3.1 is used.

3.10. Find the inverse Fourier transform of $R(fT_s)$ from Table 2.1, and sketch its curve.

3.11. Calculate the pulse-shape produced by rolling off the spectrum of $\Pi(t/T_b)$ with the function $\exp(-\pi f^2 T_s^2)$. Use the function erfc (Section 4.1.5) in your answer.

3.12. Consider the function $y(t)$ with the Fourier transform $Y(f) = R(fT_b)$. Show that $Y(f)$ satisfies (3.11), find $y(t)$, and verify that it satisfies (3.8).

3.13. Prove that the following is a decoding algorithm for the duobinary code (Section 3.3.4): if $d_k = 1$ then $b_k = 1$, and if $d_k = 0$ or 2 then $b_k = 0$.

3.14. In a Miller code an input 1 causes a transition in the output level (from + to − or vice versa) in the middle of a bit-interval. An input 0 causes a similar transition at the start of its bit interval if it follows another 0 but not if it follows a 1. Draw a diagram similar to Fig. 3.9 to show how the encoder may be represented as going from state to state, putting out suitable outputs on each transition.

3.15. Verify that if the carrier-based signal described in Section 3.4.1 is multiplied by $\sin(2\pi f_c t)$ at the receiver then the output from the low-pass filter is zero.

3.16. Show that

$$\int_\epsilon^\infty S(f)\exp(2\pi jft)\,df = \left\{\int_{-\infty}^{-\epsilon} S(f)\exp(2\pi jft)\,df\right\}^*$$

if $s(t)$ is real.

3.17. MPSK signalling is used with a raised cosine pulse-shape $R(t/T_b)$ for the phasor, where T_b is the pulse-interval. The maximum phase-change between adjacent pulses is $\Delta\theta$. Calculate the ratio of the minimum amplitude to the maximum.

4. Random variables and noise

In this chapter the mathematical repertory is increased by the addition of basic probability theory, a natural tool for the study of noise-signals. Various ideas from Chapter 3 may then be discussed in a more quantitative fashion. The basic theory is discussed in Section 4.1. This is applied to considering the properties of simple types of noise in Section 4.2. Shot-noise is analysed in some detail, both as an important example, since it is a fundamental source of noise in receivers, and also since the theory is similar to the theory of quantum noise as considered in Chapter 8. A simple model of a noisy channel, the Additive White Gaussian Noise channel or AWGN channel, is introduced in Section 4.3. It is an important model for the following reasons:

(a) It is a simple and tractable model, and the performance of codes on such a channel can be discussed in easily visualized geometric terms.

(b) It is a fairly realistic model, especially for the fundamental sources of noise, thermal noise and shot noise.

(c) It is a good model for the prestigious satellite channel.

(d) A signalling system that behaves well on such a channel may not be too bad on other types of channel as well, and so it is the standard model for analysing signalling systems.

The reliability of an AWGN channel of the polar pulsed system from Chapter 3 is assessed. The theory is then extended to consider a system of pulse-modulation which in principle can provide arbitrarily high reliability on an AWGN channel at a finite information rate. Finally, a simple model of a fading channel, the Rayleigh channel, is discussed in Section 4.4.

4.1 Probability theory

4.1.1 *Basic concepts*

The purpose of this section is to give a quick sketch of those parts of probability theory which will be needed, without any pretence at completeness or logical presentation.

Let us imagine that the result of some experiment is one of a number of discrete possible outcomes, labelled by the values of a parameter α. The set of all such outcomes is referred to as a probability space or as an *ensemble*. We suppose that we can associate a probability p_α with each outcome. These numbers must satisfy the criteria

$$p_\alpha \geq 0 \quad \text{and} \quad \Sigma_\alpha p_\alpha = 1.$$

An event is regarded as a set of outcomes. An event is said to occur if one of the outcomes making it up occurs. The probability of an event is the sum of the probabilities of the outcomes making it up. We have the following rules:

(a) Let A denote an event, and let $P(A)$ denote the probability of A. The event 'A does not occur' is called the complement of A and is made up of those outcomes not in A. We have $P(\text{'}A \text{ does not occur'}) = 1 - P(A)$. Thus the probability of correct decoding is 1 minus the probability of incorrect decoding.

(b) The probability that one of a collection of exclusive events occurs is the sum of the probabilities of each event in turn. Exclusive events have no outcomes in common and so logically cannot occur together. Thus an event and its complement are exclusive. Outcomes are always exclusive.

(c) The probability that one at least of a collection of events, not necessarily exclusive, occurs is less than or equal to the sum of the probabilities of each event in turn. This rule is often used to provide an upper bound on error probabilities and is called the *union bound*. If the sum of the probabilities is greater than or equal to 1, then the rule tells us nothing we did not know already.

(d) We now define what we mean by independence of trials. Let us consider an experiment made up of n sub-experiments or trials. For instance the experiment may consist of tossing a coin n times, and each toss is a trial. The trials are called independent if the probability of any outcome of the whole experiment is given by multiplying the probabilities of the results of each trial in turn.

We now consider some examples: let the trial consist of tossing a biased coin, with two outcomes H (head) and T (tail). Let p be the probability of H. Then the probability of T is $1 - p$.

Next let us toss the same coin four times. The overall experiment has sixteen possible outcomes, from $HHHH$ to $TTTT$. If the trials are independent, then we can find the probability of any particular outcome by multiplying together a factor p for each H and $(1 - p)$ for each T. Thus the probability of the outcome $HHTH$ (a head and another head, then a tail and finally a head) is $p^3(1-p)$.

The event '3 heads' consists of four outcomes $THHH$, $HTHH$, $HHTH$, and $HHHT$, each with probability $p^3(1-p)$. Therefore the probability of the event '3 heads' is $4p^3(1-p)$. The probability of 'not 3 heads' (an event consisting of the other twelve outcomes) is $1 - 4p^3(1-p)$. The events '3 heads' and '3 tails' are exclusive, as are the events '1 head exactly (with 3 tails)', '2 heads exactly', and '3 heads exactly'. These have probabilities $4p(1-p)^3$, $6p^2(1-p)^2$, and $4p^3(1-p)$ respectively. Hence the event '1, 2, or 3 heads' has a probability given by the sum of these values.

More generally if the experiment is made up of n independent tosses, then the probability of more than t heads is given by

$$P(\text{'more than } t \text{ heads'}) = \Sigma_{s=t+1}^{n} \binom{n}{s} p^s (1-p)^{n-s} \tag{4.1}$$

The combinatorial factor $\binom{n}{s}$ is the number of arrangements of s symbols H and $(n-s)$ symbols T and is given by

$$\binom{n}{s} = n!/\{s!(n-s)!\},$$

where $n! = 1.2.3. \ldots .(n-1).n$ if n is a positive integer, and where $0! = 1$. This result will be used later when we consider the reception of an n-bit word, with bit-errors occurring independently with probability p. We may imagine that H corresponds to a bit-error and T to a correct bit.

As an example of the application of the union bound let us consider the probability of 'at least one head' in the experiment of four tosses already discussed. This occurs whenever one (or more) of the following non-exclusive events occurs: 'H in ith position, regardless of what happens elsewhere', for $i = 1, 2, 3,$ or 4. Hence we find that P ('at least one head') $\leqslant 4p$. The exact probability may be easily and quickly calculated as 1 minus the probability of 'no heads', that is, $1 - (1-p)^4$. The union bound is useless for $p \geqslant 0.25$, but quite good for small p. Thus for $p = 0.05$ it gives an upper bound of 0.2 on the true result 0.1855.

Finally we need to discuss conditional probabilities. The probability of event A given event B is denoted by $P(A|B)$ and is equal to the probability of A and B occurring together, divided by the probability of B. Thus

$$P(A|B) = P(\text{'}A \text{ and } B\text{'})/P(B). \tag{4.2}$$

Thus in the experiment involving four tosses the probability of a head on the first toss given two heads exactly altogether is the probability of 'head in the first position and exactly one other head' divided by the probability of 'two heads exactly'.

Another example is the following: suppose that at a given moment a communications link uses one of a set of symbols s_i ($i = 1, 2, \ldots, n$) with probability $P(S_i)$, S_i being the event 's_i is used'. These events are exclusive and their probabilities add up to unity, since one of these events has to occur. Let E denote the event that an error occurs on reception. Then the events 'E and S_1', 'E and S_2', etc. are exclusive, since if they occurred together two different symbols would be sent at the same time. Moreover, E occurs if and only if one of these events occurs. Hence we find

$$P(E) = P(\text{'}E \text{ and } S_1\text{'}) + P(\text{'}E \text{ and } S_2\text{'}) + \ldots$$

$$= \Sigma_i P(\text{'}E \text{ and } S_i\text{'}).$$

Now since by (4.2)

$$P(\text{'}E \text{ and } S_i\text{'}) = P(E|S_i)P(S_i),$$

we obtain

$$P(E) = \Sigma_i P(E|S_i)P(S_i). \tag{4.3}$$

This is a useful result, giving the error probability in terms of the conditional error probabilities $P(E|S_i)$ which should be quite readily calculable, being simply the probability of error when the signal s_i is sent.

4.1.2 Random variables

Suppose that associated with each outcome α of an experiment is a 'pay-off' x_α. We can imagine this pay-off as specified by a table of numbers, one entry for each outcome. Such a table is called a discrete *random variable*. The random variable will usually be denoted by the same symbol as that denoting the entries in the table, by x in this case, but without the subscript α.

A function $f(x)$ of a random variable x is defined as the random variable with the table of values $f(x_\alpha)$. Similarly if we have another random variable y as well (also specified by a table of values y_α) then any function $f(x, y)$ of x and y is defined as the random variable with the table of values $f(x_\alpha, y_\alpha)$. Thus $x + y$ is the random variable with the table of values $x_\alpha + y_\alpha$. In consequence the usual algebraic identities like $(x + y)^2 = x^2 + 2xy + y^2$ also apply to random variables.

The mean of a random variable x is denoted by $\bar{x}$ or $\langle x \rangle$ and is defined as the number

$$\langle x \rangle = \Sigma_\alpha p_\alpha x_\alpha.$$

A constant a is independent of the outcome α and its mean is equal to itself, since $\Sigma_\alpha p_\alpha a = 1 \cdot a = a$. The mean of a random variable is a constant. The following properties of the mean are immediately verified:

$$\langle ax \rangle = a \langle x \rangle,$$

$$\langle x + y \rangle = \langle x \rangle + \langle y \rangle.$$

Generally we find that

$$\langle f(x, y) \rangle = \Sigma_\alpha p_\alpha f(x_\alpha, y_\alpha).$$

Thus the mean of the square of a random variable x is denoted by $\langle x^2 \rangle$ and is given by $\langle x^2 \rangle = \Sigma_\alpha p_\alpha x_\alpha^2$. The mean of $\langle (x-a)^2 \rangle$ is given by

$$\langle (x-a)^2 \rangle = \langle x^2 \rangle - 2a \langle x \rangle + a^2.$$

In particular the *variance* of x is given by $\langle (x-a)^2 \rangle$ with $a = \langle x \rangle$ and is denoted as var(x). Note that it is just a number, not a function of x, which would be a random variable. We have

$$\text{var}(x) = \langle (x - \langle x \rangle)^2 \rangle = \langle x^2 \rangle - \langle x \rangle^2. \tag{4.4}$$

The square-root of the variance is called the *standard deviation*, and is usually given the symbol σ. It is a measure of the spread of the random variable.

The *covariance* cov(x, y) of two random variables x and y is defined as

$$\text{cov}(x, y) = \langle (x - \langle x \rangle)(y - \langle y \rangle) \rangle. \tag{4.5}$$

It is given by

$$\text{cov}(x, y) = \Sigma_\alpha p_\alpha (x_\alpha - a)(y_\alpha - b),$$

where $a = \langle x \rangle$ and $b = \langle y \rangle$. If the covariance is zero then the random variables x and y are called *uncorrelated*. An easily proved but important result is that

$$\text{if} \quad \text{cov}(x, y) = 0 \quad \text{then} \quad \text{var}(x + y) = \text{var}(x) + \text{var}(y). \tag{4.6}$$

In most cases the random variables that we shall be dealing with have zero mean.

4.1.3 Continuous random variables

The set of all outcomes can be uncountably large, so that when we evaluate a mean we have to integrate over α rather than sum over it. Often this integration involves a multi-dimensional integral. Then it is usually more convenient to use the value of the random variable itself as the variable of integration, and it is necessary to define a *probability density* $\rho(x')$ so that the probability of finding the random variable x in the infinitesimal range x' to $x' + dx'$ is $\rho(x')\,dx'$. The function $\rho(x')$ must be non-negative, and its integral from a to b is the probability of finding x in the range a to b. In particular the integral from $-\infty$ to $+\infty$ must give unity, since the probability of x taking some value in this range is 1. Similarly with two random variables x and y it is convenient to define a two-dimensional probability density ρ_2 so that the probability of finding x in the range x' to $x' + dx'$ and of finding y in the range y' to $y' + dy'$ at the same time is $\rho_2(x', y')\,dx'\,dy'$. The function $\rho_2(x', y')$ is called the *joint probability density*. Similar definitions apply for collections of three or more random variables. The function-arguments are liable to have their names changed, and so if there is any doubt about which random variable or collection of random variables is being considered then this may be denoted by a subscript on ρ. In fact there is no need to have primes on the arguments so long as the symbol for the argument does not coincide with that denoting the random variable.

The mean of any function $f(x, y)$ of the random variables x and y is defined as

$$\langle f(x, y) \rangle = \int_{-\infty}^{\infty} \int_{-\infty}^{\infty} \rho_2(x', y') f(x', y')\,dx'\,dy'. \tag{4.7}$$

In particular the mean of x is

$$\langle x \rangle = \int_{-\infty}^{\infty} x' \rho(x')\,dx' = \int_{-\infty}^{\infty} \int_{-\infty}^{\infty} x' \rho_2(x', y')\,dx'\,dy',$$

the variance is

$$\langle (x - \langle x \rangle)^2 \rangle = \int_{-\infty}^{\infty} \rho(x')(x' - \langle x \rangle)^2\,dx', \tag{4.8}$$

and the covariance of x and y is

$$\langle (x - \langle x \rangle)(y - \langle y \rangle) \rangle = \int_{-\infty}^{\infty} \int_{-\infty}^{\infty} \rho_2(x', y')(x' - \langle x \rangle)(y' - \langle y \rangle)\,dx'\,dy'. \tag{4.9}$$

It should be noted that the means $\langle x \rangle$ and $\langle y \rangle$ are constants and that if only one random variable x appears, then we may eliminate the other using the fact that the probability density $\rho(x')$ for x and the two-dimensional probability density $\rho_2(x', y')$ are related by

$$\rho(x') = \int_{-\infty}^{\infty} \rho_2(x', y') \, dy'. \tag{4.10}$$

If the random variables x and y are the real and imaginary parts of a complex random variable z, then we may regard $\rho_2(x', y')$ as a probability density in the complex plane.

Most of the definitions and results in this section are readily generalized to situations with more than two random variables.

A set of random variables is called *independent* if the joint probability density factorizes:

$$\rho(x', y', z', \ldots) = \rho_1(x')\rho_2(y')\rho_3(z') \ldots \tag{4.11}$$

into the product of the probability density functions for each random variable. An important consequence is that the mean of a product of functions of independent random variables is the product of the means:

$$\langle f(x)g(y)h(z) \rangle = \langle f(x) \rangle \langle g(y) \rangle \langle h(z) \rangle. \tag{4.12}$$

It is easily shown that independent random variables are uncorrelated, but the converse is not necessarily true.

If all the density functions $\rho_1, \rho_2 \ldots$ in (4.11) are the same function, then the random variables are called independent identically distributed (i.i.d.) random variables.

For discrete probability distributions the probability density function becomes a sum of delta functions. Thus let the random variable X be the number of heads obtained in tossing a coin with probability p of obtaining a head. Then the probability density function of X is given by

$$\rho(x) = (1-p)\delta(x) + p\delta(x-1) \tag{4.13}$$

so that it is two sharp spikes of strengths $(1-p)$ and p respectively at $x = 0$ and $x = 1$. This approach avoids the need for separate analyses of the continuous and discrete cases.

4.1.4 The Chernoff bound

We now derive the following result due to Chernoff, which is very useful in considering the reliability of error-correcting codes in the presence of randomly occurring errors (Wozencraft and Jacobs 1965, Section 2.5). If the coin just discussed is tossed n times, the tosses being independent, then the probability that the number of heads exceeds qn, where q is a fixed number greater than p, falls *exponentially* with n. This result is proved in three stages:

PROBABILITY THEORY

(a) Let Y be a random variable with probability density function ρ_Y. Let λ be a non-negative real number and b any number. Then (provided the integrals are convergent at the upper limit, not an issue in the coin-tossing situation)

$$\langle e^{\lambda Y}\rangle = \int_{-\infty}^{\infty} \rho_Y(y) e^{\lambda y}\, dy \geq \int_{b}^{\infty} \rho_Y(y) e^{\lambda y}\, dy$$

$$\geq \int_{b}^{\infty} \rho_Y(y) e^{\lambda b}\, dy = e^{\lambda b} \int_{b}^{\infty} \rho_Y(y)\, dy. \qquad (4.14)$$

Therefore

$$P(\text{'}Y \geq b\text{'}) \leq e^{-\lambda b} \langle e^{\lambda Y}\rangle. \qquad (4.15)$$

The first inequality in (4.14) follows since the integrand is non-negative and part of the range of integration has been cut out. The second inequality follows from $e^{\lambda y} \geq e^{\lambda b}$ for $y \geq b$ (and $\lambda \geq 0$). The final integral is simply the probability of '$Y \geq b$' and hence we have (4.15). Obviously the best value of λ to choose is the one that minimizes the right-hand side.

(b) Let Y be the sum of a set of independent identically distributed random variables $X_1, \ldots, X_n$. It follows that

$$\langle e^{\lambda Y}\rangle = \langle \exp(\lambda X_1) \exp(\lambda X_2) \ldots\rangle = \langle \exp(\lambda X_1)\rangle \langle \exp(\lambda X_2)\rangle \ldots$$
$$= \{\langle \exp(\lambda X_1)\rangle\}^n.$$

First we have used the factoring property of the exponential function, then the independence property (4.12), and lastly the fact that the mean values are all the same. Hence by putting $b = nq$ in (4.15) we find

$$P(\text{'}Y \geq nq\text{'}) \leq \{\langle \exp(\lambda X_1)\rangle e^{-\lambda q}\}^n.$$

(c) We now let X_i be the number of heads on the ith toss of a coin. This random variable takes the values 1 with probability p, 0 with probability $(1-p)$. Then by (4.13) it follows that

$$\langle \exp(\lambda X_i)\rangle = (1-p) + p e^{\lambda}.$$

Thus the probability that there are more than nq heads on n tosses is less than or equal to $\{Q(\lambda)\}^n$ for all $\lambda \geq 0$, where $Q(\lambda) = \{(1-p) + p e^{\lambda}\} e^{-\lambda q}$. Obviously $Q(0)$ is equal to 1. Moreover the derivative at $\lambda = 0$ is found to be $p - q$, so that if $q > p$ the expression Q decreases (starting at 1) as λ increases from 0. So Q must achieve a value M less than 1. Hence the probability is less than or equal to M^n for some $M < 1$ and hence it falls exponentially with n.

Example. Let $p = 0.5$ and $q = 0.6$. We find $Q(0.4) = 0.9801$ as being close to the minimum. Thus the probability of 600 or more heads in 1000 tosses is less than $0.9801^{1000} < 2 \times 10^{-9}$. Similarly the probability of 6000 or more heads in 10 000 tosses is found to be less than 5.1×10^{-88}.

4.1.5 Gaussian distributions

The bell-shaped probability density

$$\rho_G(x) = (\pi\eta)^{-1/2} \exp(-x^2/\eta) \qquad (4.16)$$

for some random variable X is called a zero-mean *Gaussian* distribution, and the random variable X itself a Gaussian random variable. Here η is a fixed quantity greater than zero. If x is replaced by $x - \mu$ on the right-hand side, then the mean becomes μ. A Gaussian distribution is completely specified by its mean and variance. It is fairly simply established that

$$\int_{-\infty}^{\infty} \rho_G(x)\,dx = 1 \quad \text{and} \quad \int_{-\infty}^{\infty} x^2 \rho_G(x)\,dx = \tfrac{1}{2}\eta,$$

so that the variance is $\tfrac{1}{2}\eta$ (Exercise 4.7).

Gaussian distributions commonly arise as a consequence of the central limit theorem which in rough terms states that the sums of very many independent random variables S_i is a random variable with a distribution close to Gaussian (Chung 1974, Section 7.5). A typical example of such a random variable is the noise-voltage across a resistor produced by the thermal motion of many charged particles. Since the distribution of S is close to Gaussian it is almost completely specified by its mean and variance, and these in turn are often calculated as the sums respectively of the means and of the variances of the S_i. (Note that $\text{var}(S) = \Sigma_i \text{var}(S_i)$ since the S_i, being independent, are uncorrelated.) This is one of the reasons for the study of the variances of the noise-signals in Section 4.2.

The probability that a random variable X with probability distribution (4.16) will exceed the value x is given by

$$P('X > x') = \int_x^{\infty} \rho_G(t)\,dt = \tfrac{1}{2}\,\text{erfc}\,(x\eta^{-1/2}) \qquad (4.17)$$

where

$$\tfrac{1}{2}\,\text{erfc}\,(x) = \pi^{-1/2} \int_x^{\infty} \exp(-t^2)\,dt, \qquad (4.18)$$

and where $\tfrac{1}{2}\eta$ is the variance. (See Abramowitz and Stegun 1965, Chapter 7. 'erfc' is an abbreviation for error-function complement.) This function is a transcendental function but numerical values are easily obtained. (We shall normally keep the factor $\tfrac{1}{2}$ with the function erfc as this combination is very common.) For positive x it is closely bounded above and below through the result:

$$[x + \sqrt{(x^2 + 2)}]^{-1} < \exp(x^2) \int_x^{\infty} \exp(-t^2)\,dt \leqslant [x + \sqrt{(x^2 + 4/\pi)}]^{-1}. \qquad (4.19)$$

From it we may obtain the rather crude but none the less useful bounds

$$\tfrac{1}{2}\operatorname{erfc}(\sqrt{u}) < e^{-u}/\sqrt{(4\pi u)}, \tag{4.20}$$

$$\tfrac{1}{2}\operatorname{erfc}(\sqrt{u}) < e^{-u} \quad \text{for} \quad u \geq 0. \tag{4.21}$$

Equation (4.20) is a reasonable approximation if $u > 3$. The following result is also sometimes useful (Exercise 4.5):

$$\tfrac{1}{2}\operatorname{erfc}(\sqrt{y}) < \tfrac{1}{2}\operatorname{erfc}(\sqrt{x})\, e^{-(y-x)} \quad (\text{for } y \geq x \geq 0). \tag{4.22}$$

4.1.6 *Multivariate Guassian distributions*

A zero-mean multivariate ('many-variabled') Gaussian distribution is a collection of random variables $x_1, x_2, \ldots, x_n$ with a probability distribution

$$p(x_1', x_2', \ldots, x_n') = C \exp(-\Sigma_{kl} x_k' A_{kl} x_l')$$

where C is a normalizing constant chosen so that

$$\int_{-\infty}^{\infty} \cdots \int_{-\infty}^{\infty} dx_1'\, dx_2' \ldots dx_n'\, p(x_1', x_2', \ldots, x_n') = 1$$

and the A_{kl} are coefficients so chosen that the quadratic form $\Sigma_{kl} x_k' A_{kl} x_l'$ is always positive, except when the x_k' are all zero, when it too must be zero. This is the zero-mean distribution. The random variables are also called *mutually Gaussian*. Some useful results are:

(a) Linear combinations $y_k = \Sigma_l B_{kl} x_l$ are also random variables with a multivariate Gaussian distribution (provided that they are linearly independent). A single linear combination $\Sigma_k a_k x_k$ has a Gaussian distribution.

(b) A multivariate Gaussian distribution is completely determined by its means, variances and covariances.

(c) If $x_1, x_2, \ldots, x_n$ are uncorrelated with mean zero, then the probability density has the form

$$C \exp(-A_{11} x_1'^2 - A_{22} x_2'^2 - \ldots),$$

which obviously factorizes, so that the random variables are independent. Here the A_{ii} are all greater than zero.

4.2 Noise signals

4.2.1 *Introductory*

We model a noise-signal as a *random* function of the time t which is added to the received signal. (Such a function is sometimes called a stochastic process.) In a sense it is an infinite collection of random variables, one for each value of t. A proper treatment then involves advanced techniques. Fortunately the results we need may be derived in a straightforward way if we simply use a discrete set of functions (an 'ensemble') $x_\alpha(t)$, labelled by α, each with a probability p_α. A particular function will be called an 'example of a noise signal'. Moreover, these functions will be assumed to be periodic with a period T very much longer than any realistic time-scale in the situation. It may be helpful to imagine a physical arrangement where the noise-signal circulates indefinitely round a loop of wave-guide or coaxial cable, repeating itself after every interval T (Section 2.3.5). In particular cases an 'ensemble' average is replaced by a process of integration, usually with a multi-dimensional integral.

4.2.2 The autocorrelation function

Because of the periodicity each function $x_\alpha(t)$ can be expanded as a Fourier series

$$x_\alpha(t) = \Sigma_n X_{\alpha n} \exp(2\pi j nt/T). \qquad (4.23)$$

We impose two conditions. The first is simply the zero-mean condition,

$$\langle x(t) \rangle = \Sigma_\alpha p_\alpha x_\alpha(t) = 0.$$

It follows that

$$\langle x(t) \rangle = \Sigma_n \langle X_n \rangle \exp(2\pi j nt/T) = 0 \quad \text{for all } t.$$

Hence the uniqueness result of Section 2.3.1 gives

$$\langle X_n \rangle = 0 \quad \text{for all } n. \qquad (4.24)$$

The second condition is the much stronger 'stationary condition', that

$$\langle x^*(\tau) x(\tau + t) \rangle \quad \text{is independent of } \tau \text{ for all } t.$$

(The complex conjugation is not necessary for real signals, but nevertheless is a convenient guide in the subsequent manipulations.) Hence we may define the autocorrelation function

$$\tilde{a}(t) = \langle x^*(\tau) x(\tau + t) \rangle \qquad (4.25)$$

which by hypothesis is independent of τ for all t. In particular we note that $\tilde{a}(0)$ is the mean power, which must be time-independent. Since the functions $x_\alpha(t)$ are periodic with period T, so is $\tilde{a}(t)$. The tilde emphasizes this. By expanding the factors on the right-hand side of (4.25) as Fourier series we find (with the sums from $-\infty$ to $+\infty$)

$$\tilde{a}(t) = \Sigma_{mn} \exp\{2\pi j(n-m)\tau/T\} \exp(2\pi j nt/T) \langle X_m^* X_n \rangle$$
$$= \Sigma_p \exp(2\pi j p\tau/T) [\Sigma_n \exp(2\pi j nt/T) \langle X_{n-p}^* X_n \rangle]. \qquad (4.26)$$

Here we have made the sum over m the inside sum, replaced m by $p = n - m$

NOISE SIGNALS

(still with limits $-\infty$ to $+\infty$), and then interchanged the order of the sums over p and n, making the sum over n the inside sum. Since this expression is independent of τ only the terms with $p = 0$ survives, and so the factor in square brackets must vanish if $p \neq 0$. Since this is true for all t it follows that $\langle X^*_{n-p} X_n \rangle = 0$ for all n if $p \neq 0$. Hence we find that

$$\langle X^*_m X_n \rangle = 0 \qquad \text{for } m \neq n. \tag{4.27}$$

Hence by (4.26) it follows that

$$\tilde{a}(t) = \Sigma_n \exp(2\pi j n t / T) \langle X^*_n X_n \rangle \tag{4.28}$$

and hence we may identify the Fourier coefficients A_n of $\tilde{a}(t)$ as $A_n = \langle X^*_n X_n \rangle$. Conversely we may write

$$\langle X^*_m X_n \rangle = A_n \delta_{mn}. \tag{4.29}$$

Here δ_{mn} is the simple but useful Kronecker symbol defined by

$$\delta_{mn} = \begin{cases} 1 & \text{if } m = n, \\ 0 & \text{if } m \neq n. \end{cases}$$

It should be noted that since A_n is the mean of a square modulus it must be real and non-negative, in consequence of which we find $\tilde{a}(-t) = \tilde{a}(t)$.

The function $\tilde{a}(t)$ may be regarded as the correlation of two random variables $x(\tau)$ and $x(\tau + t)$ which are the noise-signal amplitudes at two specified instants separated by t. We would then expect $\tilde{a}(t)$ to fall to zero as $|t|$ becomes larger than some 'memory time' T_M, since the amplitudes would then become uncorrelated. Of course $\tilde{a}(t)$ has an artificially imposed periodicity with period T, but T is always chosen much larger than any other time-scales including T_M. Hence we may write $\tilde{a}(t)$ as the periodic repetition of the true autocorrelation function $a(t)$ which goes to zero rapidly as $|t|$ gets much larger than T_M:

$$\tilde{a}(t) = \Sigma_k a(t - kT). \tag{4.30}$$

Thus for $|t|$ less than, say, $T/2$ we may approximate $\tilde{a}(t)$ by $a(t)$, since the repeats in (4.30) are displaced by multiples of T and hence are negligible. The result becomes exact in the limit of large T. In particular the mean noise-power N is $a(0)$ with negligible error and hence is given by

$$N = a(0) = \int_{-\infty}^{\infty} A(f) \, df. \tag{4.31}$$

In fact $A(f)$ gives the distribution of the power over the various frequencies. It is called the two-sided *power spectral density* and is measured in units of watts/hertz or joules. Because of (4.30) the Fourier coefficients of $\tilde{a}(t)$ are given in terms of the Fourier transform $A(f)$ of $a(t)$ as

$$A_n = T^{-1}A(n/T) \qquad (4.32)$$

as shown in Section 2.3.4.

An important consequence of (4.29) is that the real and imaginary parts of X_n are uncorrelated, if the noise $x(t)$ is real. For then we find, using $X^*_{-m} = X_m$, that for $m, n > 0$

$$\langle X_m X_n \rangle = \langle X^*_{-m} X_n \rangle = 0 \qquad (4.33)$$

by (4.29), since with $m, n > 0$ we cannot have $m = -n$. Adding this to (4.29) gives

$$\langle \operatorname{Re}(X_m) X_n \rangle = \tfrac{1}{2} A_n \delta_{mn},$$

and so by taking the real and imaginary parts and by using the reality of A_n we find

$$\langle \operatorname{Re}(X_m) \operatorname{Re}(X_n) \rangle = \tfrac{1}{2} A_n \delta_{mn}, \qquad (4.34a)$$

$$\langle \operatorname{Im}(X_m) \operatorname{Re}(X_n) \rangle = 0. \qquad (4.34b)$$

Similarly by subtracting (4.33) from (4.29) we find

$$\langle \operatorname{Im}(X_m) \operatorname{Im}(X_n) \rangle = \tfrac{1}{2} A_n \delta_{mn}. \qquad (4.34c)$$

4.2.3 The effect of filtering on noise

If a noise signal $x_\alpha(t)$ is put through a filter with impulse-response function $h(t)$, then the output $y_\alpha(t)$ is the convolution of h with x_α. Thus we find by (2.49)

$$\langle Y^*_m Y_n \rangle = H^*(m/T)H(n/T)\langle X^*_m X_n \rangle = H^*(n/T)H(n/T)A_n \delta_{mn}.$$

We have replaced m by n in the first factor $H^*(m/T)$, since the whole term vanishes in any case if $m \neq n$. Thus we may write

$$\langle Y^*_m Y_n \rangle = T^{-1} A'(n/T) \delta_{mn}$$

with

$$A'(f) = |H(f)|^2 A(f). \qquad (4.35)$$

The inverse Fourier transform $a'(t)$ of $A'(f)$ is of course the true autocorrelation function of the filter output $y(t)$.

4.2.4 White noise

'White noise' is noise where $A(f)$ is a constant, at least in the frequency range of interest. $A(f)$ is usually specified as

$$A(f) = \tfrac{1}{2}\eta, \qquad (4.36)$$

NOISE SIGNALS

where η is a positive constant called the *one-sided noise spectral density*. Like $A(f)$ it is measured in units of energy. Thus the real output $y(t)$ from a filter with transfer function $H(f)$ when white noise is put in has a power spectral density given by

$$A'(f) = \tfrac{1}{2}\eta H^*(f)H(f).$$

Thus the autocorrelation function is proportional to the convolution of the real impulse response function $h(t)$ with $h^*(-t) = h(-t)$, and so it is given by

$$\langle y(\tau)y(\tau+t)\rangle = a'(t) = \tfrac{1}{2}\eta \int_{-\infty}^{\infty} h(t')h(t'-t)\,dt'. \qquad (4.37)$$

It is independent of τ (Section 4.2.2).

Suppose white noise is passed through a band-pass filter with bandwidth W, with the transfer function $H(f)$ equal to zero unless $|f|$ lies in some range of width W, when it is unity. The characteristic is shown in Fig. 2.4(b). Then the power spectral density is equal to $\tfrac{1}{2}\eta$ in two ranges of width W, one on the positive-frequency side of zero and the other on the negative-frequency side. It is zero elsewhere. Hence the mean power as given by (4.31) is equal to $\tfrac{1}{2}\eta \cdot 2W = \eta W$. Thus the noise power is proportional to the bandwidth, with η as the constant of proportionality. An example of white noise passed through a band-pass filter is shown in Fig. 4.1. Here the bandwidth W is one-third of the carrier frequency f_c. The ticks on the x-axis are put in at twice the rate determined by the carrier frequency to illustrate how the phase varies randomly, as well as the amplitude.

An expression for the correlation of the Fourier coefficients of white noise

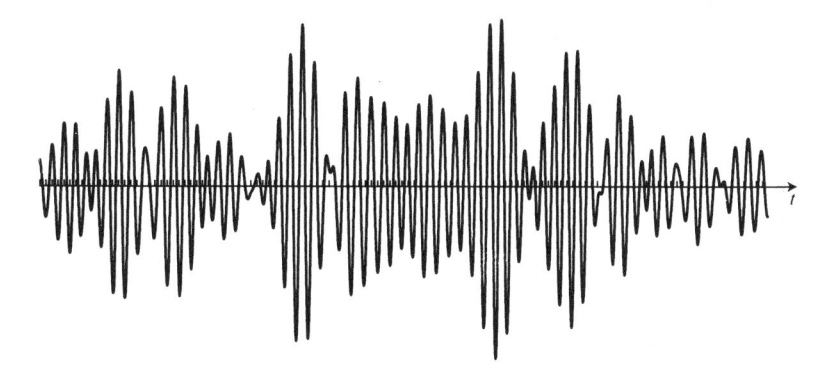

FIG. 4.1 Simulated Gaussian white noise passed through a band-pass filter with the transfer function of Fig. 2.4(b) with $W/f_c = 0.3333$. The ticks on the axis, spaced by the time $(2f_c)^{-1}$, show that the phase as well as the amplitude varies randomly.

forced to be periodic is readily derived. By (4.29), (4.32), and (4.36) it is found that

$$\langle X_m^* X_n \rangle = A_n \delta_{mn} = \tfrac{1}{2} \eta T^{-1} \delta_{mn}. \tag{4.38}$$

One cause of white noise is the thermal agitation of the charged carriers in the front end of a receiver, or the thermal agitation of the electromagnetic field itself (black-body radiation). It can be shown that in this case $\eta = k_B \theta_N$, where θ_N is the temperature and k_B is Boltzmann's constant 1.38×10^{-23} J/K, which at room temperature $\theta_N = 290$ K is 4×10^{-21} J. Thus the noise-power at the input of a television receiver due to thermal noise would be about 4×10^{-14} W if the bandwidth W were 10 MHz.

The strength of thermal noise is readily derived by using the equipartition theorem of statistical mechanics (Reif 1965) on the loop-model of Section 2.3.5. Let us consider a particular normal mode, the nth. This mode has two degrees of freedom, the in-phase and quadrature amplitudes, and so by the equipartition theorem its mean energy is $k_B \theta_N$, where θ_N is the temperature of the heat-bath keeping the radiation in the loop warm. As was shown in Section 2.3.5 the energy of the nth mode is $2T X_n^* X_n$, and so its complex amplitude satisfies

$$2T \langle X_n^* X_n \rangle = k_B \theta_N.$$

Hence by (4.38) we find $\eta = k_B \theta_N$.

4.2.5 Shot noise

As an example of practical importance we consider another type of noise called *shot noise* which is produced in electronic devices where the electrons have to cross a barrier, doing so at random times. The average current through the device is determined by the average value λ of the number of electrons crossing each second. The actual current consists of a stream of pulses occurring independently of one another at random times. (Such a process is known as a 'Poisson process', with intensity λ.) Let $h(t - \tau)$ be the current produced by a carrier which starts to cross at time τ. This current is zero for $t < \tau$, remains finite while the carrier is moving across the barrier, and then drops back to zero when it has reached the other side, or more generally when the electrical disturbance produced by the crossing has died away. Thus $h(t)$ is a localized function with some time-scale T_c, the crossing time. We may regard the current pulse $h(t - \tau)$ as though it were the output of a filter with an impulse-response function $h(t)$ driven by a δ-pulse at $t = \tau$. Any subsequent filtering can then be taken into account by including it with this hypothetical filter, so that $h(t)$ becomes the impulse-response function of a filter consisting of the hypothetical filter followed by the actual filter. The overall current is thus the sum of a succession of impulses-responses which may or may not overlap significantly. These impulses occur at a mean rate λ every second, although in a random manner.

NOISE SIGNALS

We model the situation as follows: to avoid analytical difficulties we consider only those crossings which start in a long interval $-T$ to $+T$. (Periodic boundary conditions are not imposed here.) This interval is divided into $(2N + 1)$ equal subintervals of duration $\Delta = 2T/(2N + 1)$, labelled from $-N$ to $+N$. The duration Δ is very much less than both T_c and λ^{-1}, the mean time between the starts of crossings. (At the end we take the limit $\Delta \to 0$, $N \to \infty$ with T fixed, and then the limit $T \to \infty$.) We next assume that the probability of a crossing starting in a given subinterval is $\lambda \Delta$, a value very much less than unity by our choice of Δ. Thus at least to first order the probability of a crossing starting in a given subinterval is proportional to the length of that subinterval. The chances of two or more crossings starting in the same subinterval are negligible. Moreover, it will be assumed that this probability is independent of what happens in the other subintervals. Thus the situation may be simulated as follows: a very heavily biased coin is tossed $(2N + 1)$ times, these trials being independent (Section 4.1.1). The probability of a head on any toss is the small number $\lambda \Delta$. Thus there are altogether $2^{(2N+1)}$ possible outcomes α, not of course all with the same probability. We then set the random variable ν_k as the number of heads on the kth toss. (In Section 4.1.4 this random variable was denoted by X_k.) Thus $\nu_{\alpha k} = 1$ if the outcome α has a head in the kth location, 0 if not. Then we note that:

(a) The mean number of heads on the kth toss is

$$\bar{\nu}_k = \langle \nu_k \rangle = \lambda \Delta \cdot 1 + (1 - \lambda \Delta) \cdot 0 = \lambda \Delta. \tag{4.39a}$$

(b) The covariance satisfies

$$\langle (\nu_k - \bar{\nu}_k)(\nu_l - \bar{\nu}_l) \rangle = \langle \nu_k \nu_l \rangle - \bar{\nu}_k \bar{\nu}_l = \lambda \Delta (1 - \lambda \Delta) \delta_{kl}, \tag{4.39b}$$

where $\delta_{kl} = 1$ if $k = l$, $= 0$ if $k \neq l$. The middle form is obtained by expanding the left-hand side and remembering that $\bar{\nu}_k$ and $\bar{\nu}_l$ are constants which can be taken outside the averaging brackets. If $k \neq l$ then by a result analogous to (4.12) we have $\langle \nu_k \nu_l \rangle = \langle \nu_k \rangle \langle \nu_l \rangle$ since ν_k and ν_l are independent random variables. Hence the covariance is zero. We also find that $\langle \nu_k^2 \rangle = \lambda \Delta \cdot 1^2 + (1 - \lambda \Delta) \cdot 0^2 = \lambda \Delta$, and hence the result for $k = l$.

We now reconsider the shot-noise problem. The current is the random function of time

$$s_\alpha(t) = \Sigma_{k=-N}^{N} \nu_{\alpha k} h(t - k\Delta). \tag{4.40}$$

The mean current is

$$\bar{s}(t) = \langle s(t) \rangle = \Sigma_{k=-N}^{N} \langle \nu_k \rangle h(t - k\Delta) = \lambda \Delta \Sigma_{k=-N}^{N} h(t - k\Delta). \tag{4.41}$$

We next consider the noise as the deviation from the mean:

$$x_\alpha(t) = s_\alpha(t) - \bar{s}(t).$$

Then we find

$$\langle x(t)^*x(t')\rangle = \Sigma_{k=-N}^{N}\Sigma_{l=-N}^{N}\langle(\nu_k - \bar{\nu}_k)(\nu_l - \bar{\nu}_l)\rangle h(t - k\Delta)^*h(t' - l\Delta).$$

(The complex conjugation is introduced only to match this expression with similar expressions introduced previously. The $x(t)$ are of course real.) After the substitution of (4.39b) into this equation the factor δ_{kl} removes all terms with $l \neq k$ to leave a single summation:

$$\langle x(\tau)^*x(\tau + t)\rangle = \lambda\Delta(1 - \lambda\Delta)\Sigma_{k=-N}^{N}h(t - k\Delta)^*h(t + \tau - k\Delta). \quad (4.42)$$

We now take the limit $\Delta \to 0$, $N \to \infty$, keeping T fixed. Then the sum becomes a Riemann integral. Finally we let $T \to \infty$. We do this first with (4.41) and obtain

$$\langle s(t)\rangle = \lambda\int_{-\infty}^{\infty} h(t - \tau)\,d\tau,$$

where $d\tau$ has replaced Δ. Setting $\tau' = t - \tau$ gives

$$\langle s(t)\rangle = \lambda\int_{-\infty}^{\infty} h(\tau')\,d\tau' = \lambda H(0), \quad (4.43)$$

where $H(f)$ is the Fourier transform of $h(t)$. Evidently $\langle s(t)\rangle$ is independent of t. The same limiting processes on (4.42) give

$$\langle x(\tau)^*x(\tau + t)\rangle = \lambda\int_{-\infty}^{\infty} h(\tau - t')^*h(\tau + t - t')\,dt'.$$

We set $t'' = \tau + t - t'$ and use the reality of $h(t)$ to obtain

$$\langle x(\tau)^*x(\tau + t)\rangle = \lambda\int_{-\infty}^{\infty} h(t'')h(t'' - t)\,dt'',$$

a result very similar in form to (4.37). The left-hand side is the autocorrelation function $a(t)$, which is independent of τ, since τ does not appear on the right-hand side. If we take the Fourier transform we obtain

$$A(f) = \lambda|H(f)|^2 \quad (4.44)$$

where $A(f)$ is the Fourier transform of $a(t) = \langle x(\tau)^*x(t + \tau)\rangle$ and is thus the noise spectral density (Section 4.2.2). The Fourier transform $H(f)$ of $h(t)$ would be expected to 'roll off' at a frequency of the order of T_c^{-1}, where T_c is the crossing time.

Let us now estimate the strength of shot-noise. We suppose that we are working at frequencies well below T_c^{-1}, so that $|H(f)|$ can be approximated by its real value at zero frequency. We set

$$A(f) \approx \lambda\{H(0)\}^2. \quad (4.45)$$

As in (4.36) we define an effective one-sided noise spectral density η_s by

$$\eta_s = 2A(f). \quad (4.46)$$

Thus the noise power through a filter of bandwidth W (Section 4.2.4) is given by

$$N = \eta_s W. \tag{4.47}$$

We take the signal power as

$$S = \bar{s}^2 = \lambda^2 \{H(0)\}^2 \tag{4.48}$$

by (4.43) and hence the signal-to-noise ratio is given by

$$S/N = \lambda/(2W).$$

If we wish to estimate η_s we have to watch our units. We have called $\bar{s}$ the current, but within the conventions of this book $\bar{s}^2$ has the units of power. Thus we must set

$$\bar{s}^2 = \bar{i}^2 R \tag{4.49}$$

where $\bar{i}$ is the true electric current and R the 'load resistance' connected to the output of the device. The crossing rate λ is related to the mean electric current $\bar{i}$ by

$$\bar{i} = \lambda q, \tag{4.50}$$

where q is the carrier-charge. Hence we find by (4.46), (4.45), (4.43), (4.50), and (4.49) that

$$\eta_s = 2q\bar{i}R = 2q\bar{v} \tag{4.51}$$

where $\bar{v} = \bar{i}R$, the voltage produced across the load resistance by the mean current.

4.2.6 Gaussian noise

An important model for a noise-signal is called Gaussian noise, in which it is assumed that the random variables $\text{Re}(X_n)$ and $\text{Im}(X_n)$ for $m, n > 0$ have a multivariate Gaussian distribution with mean zero. Thus from Section 4.1.6 it follows that

(a) These random variables are independent, since according to (4.34) they are uncorrelated, and
(b) the linear combination $x(t)$ given by (4.23) must also have a Gaussian distribution.

A typical source of Gaussian noise is thermal noise, being the contribution of very many small independent sources. Something like Gaussian noise is also produced if shot noise is put through a low-pass filter or through a narrow-band band-pass filter with an impulse-response $h(t)$ that lasts for a time long in comparison with the mean time between pulses. Then at any fixed moment the noise-signal is the sum of a large number of tails of response functions starting at various times. Hence it is the sum of a large number of independent random variables, and thus will have a distribution close to Gaussian.

4.3 Gaussian noise in polar signalling

4.3.1 *Introductory*

We now consider an important special case which shows how noise causes errors in a signalling system. We suppose that a polar rectangular pulse-train is being transmitted, and that on its way to the receiver some Gaussian white noise is added. Such a transmission path will be referred to as an Additive White Gaussian Noise (AWGN) channel. Let us briefly consider these assumed properties of the noise in turn: First of all the noise is added to the signal. This then excludes cases where the noise modulates or multiplies the signal, as in a fading channel, or as in a situation where brief disconnections occur at random. Secondly the noise is white, which means that it has a uniform spectral density in the band of frequencies of interest. Thirdly the noise is Gaussian. This assumption is used towards the end of the argument, when a probability density, rather than just a variance, is needed to determine the error-probability. This excludes impulsive noises, like the clicks and splutterings that plague telephone channels.

The signalling protocol is agreed to in advance and we assume that the receiver has been synchronized so that the times at which the pulses commence and stop are known.

4.3.2 *Integrating a single pulse*

To start with we need consider only a single rectangular pulse. Thus we take the signal $s(t)$ (without the noise) as a rectangular pulse

$$s(t) = \begin{cases} c & \text{if } 0 < t < T_b, \\ 0 & \text{otherwise.} \end{cases}$$

To this is added a stationary white-noise signal $x(t)$ with one-sided spectral density η, so that an example of a received signal is $r_\alpha(t) = s(t) + x_\alpha(t)$. Without the noise the pulse-height c may be determined by sampling $r(t)$ at any moment within the pulse, but the noise makes such a measurement unreliable. A better technique is to average $r(t)$ over the whole interval 0 to T_b (Section 3.2.1). In fact this is the best that can be done, a point discussed in the next chapter. This may be achieved by passing the signal through a sliding-average filter (Section 2.2.6) with an impulse-response function

$$h(t) = \begin{cases} T_b^{-1} & \text{if } 0 < t < T_b, \\ 0 & \text{otherwise,} \end{cases}$$

and then sampling the output from the filter at $t = T_b$. The sample output is the sum of the average of $s(t)$ and of the filtered noise $y(t)$ at $t = T_b$. The first term is just c and the second term has a variance $\langle y(T_b)y(T_b)\rangle$. This is evaluated by setting $t = 0$ in (4.37) and by using the above specification of $h(t)$. After a trivial integration the result is

$$\langle y(T_b)y(T_b)\rangle = a'(0) = \tfrac{1}{2}\eta T_b^{-1}.$$

4.3.3 Error probability

The output from the sampler is a real number $c + y_\alpha(T_b)$ and the next stage in the receiver is to interpret this as '+' or '−', in an attempt to reconstruct the bit which this particular pulse conveys. The simplest decision procedure is to make a hard decision, where the choice is determined by the sign of the output.

We assume that a negative pulse with total energy E is sent, representing '−', Then the amplitude c is equal to $-\sqrt{(E/T_b)}$, since

$$\int_{-\infty}^{\infty} s^2(t)\,dt = E.$$

An error occurs if $y(T_b) > \sqrt{(E/T_b)}$, because in that case $c + y(T_b)$ is positive, and so '+' is output when '−' was sent. If the noise is Gaussian then $y(T_b)$ is a Gaussian random variable and we find from (4.17) that the error probability is given by

$$P_e = \tfrac{1}{2}\,\text{erfc}\{\sqrt{(E/\eta)}\}, \tag{4.52}$$

since in contrast to the situation in Section 4.1.5 the variance is not $\tfrac{1}{2}\eta$ but $\tfrac{1}{2}\eta T_b^{-1}$. By symmetry we must have the same probability of error for positive pulses.

The error probability has been discussed in terms of a single pulse but it is evident that our result applies to each pulse in turn in a pulse-train. In terms of the signal power P we have $E = PT_b$, so that the error probability P_e is given by

$$P_e = \tfrac{1}{2}\,\text{erfc}\{\sqrt{(PT_b/\eta)}\} \approx \exp(-PT_b/\eta). \tag{4.53}$$

Thus we find that $-\ln P_e \approx PT_b/\eta$, while the information rate J in bits/s is evidently T_b^{-1}.

We note that on channels with Gaussian noise the error-probability falls exponentially with the signal-power P. On other types of channel an increase in the power is not nearly so effective. Thus if the errors are caused by huge bursts of interference in the vicinity of the receiver, then astronomically large power levels at a remote transmitter are needed to overcome the effects of this interference.

4.3.4 Reliable communication

What does the result (4.53) imply about reliable communication? In order to make $-\ln P_e$ very large for fixed η we must increase P or increase T_b. The first alternative may not be practicable and the second slows down the information rate. We have to let this rate tend to zero in order to obtain complete reliability. Shannon showed, however, that in principle it was possible with fixed values of P and η to attain arbitrarily high reliability without the information rate falling to zero (Section 5.3). This is usually done by the use of error-correcting codes (Chapters 6 and 7).

To show that reliable communication is possible in principle on the AWGN channel we employ the following system (which is of a type known as Pulse Position Modulation or PPM). Let us imagine that the interval T_W used for transmitting each 'word' is divided into M subintervals or 'slots'. The transmitter is quiescent in all but one of these slots in which a huge rectangular pulse is sent. Thus there are M possible words. The energy of this pulse at the receiver is given by

$$E = PT_W \qquad (4.54)$$

where P is the mean received power and its amplitude is $\sqrt{(E/T_b)}$, where

$$T_b = T_W/M \qquad (4.55)$$

is the pulse-duration. The information conveyed by choosing one of M words is $\log_2 M$ bits. Hence the information-rate J is given in terms of the scaled rate

$$J' = J \ln 2 \qquad (4.56)$$

by

$$J'T_W = \ln M. \qquad (4.57)$$

The receiver attempts to find out which slot is pulsed. We suppose that it measures the amplitude in each slot. A threshold is set at $\frac{1}{2}\sqrt{(E/T_b)}$, that is at half the pulse-amplitude. If only one amplitude exceeds this threshold then that is taken as denoting the pulsed slot. An error or decoding failure occurs if the amplitude in the true slot is forced down by the noise below the threshold or if the noise forces up the amplitude in another slot above the threshold. The probability for any one of these events is given by $\frac{1}{2}\text{erfc}\{\frac{1}{2}\sqrt{(E/\eta)}\}$, by an argument like the one for obtaining the error-probability in polar signalling (Section 4.3.3). The factor $\frac{1}{2}$ in the argument of the error-function arises because the noise has to exceed a threshold $\frac{1}{2}\sqrt{(E/T_b)}$, rather than $\sqrt{(E/T_b)}$. According to the union bound (Section 4.1.1) the overall 'word-error' probability P_W is always less than or equal to the sum of the probabilities of the individual events, the occurrence of any of which gives an error. Thus we find

$$P_W < M \cdot \tfrac{1}{2} \text{erfc}\{\tfrac{1}{2}\sqrt{(E/\eta)}\} < M \cdot \exp(-\tfrac{1}{4}PT_W/\eta)$$

(using (4.21)) and hence it follows that

$$\ln P_W < -\tfrac{1}{4}PT_W/\eta + \ln M = -T_W(\tfrac{1}{4}P/\eta - J'). \qquad (4.58)$$

by (4.57). Thus provided that the scaled rate $J' = J \ln 2$ is set lower than $\frac{1}{4}P/\eta$ we see that the error probability falls exponentially with T_W as T_W grows very large, without our having to arrange that J falls to zero or that P tends to infinity.

The threshold has been set at half the pulse-amplitude to make the calculation simple, and it is not by any means an optimal setting (Exercise 4.14). A better strategy is simply to use the largest amplitude to determine the pulsed slot, and then it may be shown that reliable communication in the sense just described may be achieved for large T_W provided that $J' < P/\eta$, in which case again the error-probability falls exponentially with T_W (Section 5.2.4).

Unfortunately letting T_W grow large with J, P, and η fixed soon leads to trouble:

(a) The pulse-amplitude is

$$\sqrt{(E/T_b)} = \sqrt{P} \cdot \exp(\tfrac{1}{2}J'T_W)$$

by (4.54), (4.55), and (4.57). It thus grows exponentially with T_W.

(b) The bandwidth is of the order of T_b^{-1}. Since

$$T_b/T_W = M^{-1} = \exp(-J'T_W)$$

by (4.55) and (4.57) we find that T_b^{-1} grows roughly exponentially with T_W.

(c) The amount of computing (the computing complexity) is roughly M comparisons for each decoding, and this also grows exponentially with T_W, since $M = \exp(J'T_W)$ by (4.57). If we set $P_W = \exp(-\alpha T_W)$, assuming (correctly) that the error-probability does fall exponentially with T_W provided $J' < P/\eta$, then we find that

$$P_W = M^{-\alpha/J}. \tag{4.59}$$

Thus the error probability falls off quite slowly as a power, in practice rather a low power, of the computing complexity.

4.4 Rayleigh fading

4.4.1 *Rayleigh channel*

Suppose that the bipolar signal of the last section is sent by BPSK modulation over a radio channel which is subject to fading. The Rayleigh model is a simple model to describe this fading. The received signal arrives as the sum of many signals travelling by slightly different paths, whose characteristics vary slowly in time on a time-scale T_F, the fading time. The received carrier may be represented in the phasor representation as

$$s_c(t) = \sqrt{\tfrac{1}{2}} \cdot \exp(2\pi j f_c t)\phi(t) + \sqrt{\tfrac{1}{2}} \cdot \exp(-2\pi j f_c t)\phi^*(t)$$

where f_c is the carrier-frequency and the complex phasor $\phi(t)$ is the sum of many independent slowly varying complex random variables. Thus it is reasonable to model $\phi(t)$ as a complex random function with (at any instant) a bivariate ('two-variabled') Gaussian distribution in the complex plane, so that the probability of finding Re ϕ at that moment in the range x to $x + dx$ and Im ϕ at the same moment in the range y to $y + dy$ is $\rho(x, y) \, dx \, dy$ with

$$\rho(x, y) = (\pi P)^{-1} \exp\{-(x^2 + y^2)/P\}.$$

Here P is the mean power $\langle \phi^*(\tau)\phi(\tau) \rangle$, which is independent of τ. The autocorrelation function $\langle \phi^*(\tau)\phi(\tau + t) \rangle$ falls off to zero as $|t|$ exceeds the time-scale T_F.

4.4.2 Error probability

The probability of finding $|\phi|$ in the range r to $r + dr$ is easily calculated by integrating $\rho(x, y)$ over an annulus of radius r and thickness dr. It is simply

$$P('r < |\phi| < r + dr') = 2rP^{-1} \exp(-r^2/P) \, dr.$$

We now imagine that stationary white noise is added to the signal just before it is received but after it has been modulated by the fading. We shall assume that coherent demodulation is used at the receiver to restore the polar signal, and that somehow the local oscillator can track the phase $\arg(\phi)$ of the received signal, even though it may swing round quite violently if $\phi(t)$ passes close to 0, just when it is most obscured by the additive noise. It will also be assumed that the fading time T_F is very much longer than the pulse-width T_b. Errors are more likely to occur when $|\phi|$ is small, and since $|\phi|$ is liable to remain small for a period of the order of T_F which encompasses many pulses, the errors are likely to occur in bursts. None the less in spite of the fact that the mean error probability does not give the whole story we now proceed to calculate it. According to (4.53) the error probability in receiving a pulse is $\frac{1}{2}\text{erfc}\{\sqrt{(|\phi|^2 T_b/\eta)}\}$ with the power P replaced by $|\phi|^2$. Thus we find that the error probability is given by

$$P_e = \int_0^\infty 2rP^{-1} \exp(-r^2/P) \tfrac{1}{2}\text{erfc}\{r\sqrt{(T_b/\eta)}\} \, dr. \qquad (4.60)$$

This integral is readily evaluated using integration by parts to give

$$P_e = \tfrac{1}{2}[1 - \sqrt{\{\gamma/(\gamma+1)\}}] = \tfrac{1}{2}[1 + \gamma + \sqrt{\{\gamma(\gamma+1)\}}]^{-1}, \qquad (4.61)$$

with $\gamma = PT_b/\eta$.

Although this model is a serious oversimplification of what actually happens on a fading channel it is worth making two points:

(a) Tracking the phase of the received carrier is not just a matter of maintaining accurate timing at the receiver. Modulation systems like DPSK or FSK which do not depend on accurate tracking of the carrier-phase are better on fading channels, although they do not look so good as some other coherent systems on AWGN channels.

RAYLEIGH FADING

(b) Increasing the power P is not a very effective way to reduce the error probability which falls of as P^{-1} for large P, rather than exponentially as on an AWGN channel. Thus for reliable communications some form of error-correcting coding, preferably with the ability to correct errors that occur in bursts, is necessary.

For further details on the theory of fading channels see Schwartz, Bennett, and Stein (1966), Chapter 9.

For further reading: The books by Chung (1974), Hoel, Port, and Stone (1971), and Parzen (1960) are well-known introductory texts on probability theory.

Exercises

4.1. The combinatorial factor $\binom{n}{s}$ is the number of arrangements of s symbols H and $(n-s)$ symbols T. Prove that it is equal to

$$\binom{n}{s} = n!/\{s!(n-s)!\},$$

where $n! = 1.2.3.\ldots.(n-1).n$ if n is a positive integer, and where $0! = 1$.

4.2. Is (4.12) generally true for uncorrelated random variables, which may not be independent? Find a counter-example if your answer is no.

4.3. Find an expression like (4.13) for the probability-density function of a discrete random variable that takes the value x_i with probability p_i.

4.4. By considering $\langle (Y-a)^2 \rangle$ show that

$$P(`|Y-a| \geqslant b') \leqslant \langle (Y-a)^2 \rangle / b^2.$$

Hence show that if Y is the sum of n independent identically distributed random variables $X_1, \ldots, X_n$ each with mean μ and variance σ^2 then for any fixed positive value η

$$P(`|Y/n - \mu| \geqslant \eta') \leqslant \sigma^2 \eta^{-2} n^{-1}$$

so that this probability tends to zero as n tends to infinity. Put this result into words.

4.5. With $I(x) = \tfrac{1}{2} \operatorname{erfc}(\sqrt{x})$ show that

$$I(y) \leqslant e^{-(y-x)} I(x) \quad \text{for } y \geqslant x \geqslant 0,$$

using a suitable substitution in (4.18).

4.6. Derive the inequality (4.20) from the definition (4.18).

4.7. Show that for any random variable X

$$\langle e^{Xt} \rangle = \sum_{n=0}^{\infty} \langle X^n \rangle t^n / n!.$$

Hence find $\langle X^n \rangle$ if X has the Gaussian probability density function ρ_G (4.16) by evaluating

$$\int_{-\infty}^{\infty} \rho_G(x) e^{xt} dx$$

and expanding the answer as a power series in t.

4.8. Use the Central Limit Theorem and eqn (4.20) to find the probability of 600 or more heads in 1000 independent tosses of a coin for which the probability of a head on any toss is $\frac{1}{2}$. Do the same for 6000 or more heads in 10 000 tosses, and compare your answers with those of the example at the end of Section 4.1.4. (Hint: The approximately Gaussian distribution for the total number of heads must have the correct mean and variance.)

4.9. Let A_N be the surface 'area' of a unit hypersphere in N dimensions. Show that

$$\left\{ \int_{-\infty}^{\infty} \exp(-x^2) dx \right\}^N$$

is $\pi^{N/2}$ on the one hand and

$$\int_0^{\infty} A_N r^{N-1} \exp(-r^2) dr$$

on the other hand. Hence show that

$$A_N = 2\pi^{N/2}/(N/2 - 1)!$$

by using

$$n! = \int_0^{\infty} u^n e^{-u} du.$$

4.10. Show that $\bar{a}(t)$ attains its maximum value for $t = 0$. (Hint: Use $\langle \{x(t) - x(t+\tau)\}^2 \rangle \geq 0$ and the 'stationary condition'.)

4.11. Extend the results (4.34) to the case when one or both of m and n is zero.

4.12. White noise with one-sided spectral density η is fed into a filter with impulse-response function $h(t) = e^{-at}$ for $t > 0$, $h(t) = 0$ for $t \leq 0$. Find the mean output power.

4.13. Show that if the energy of each normal mode in the loop of Section 2.3.5 is $k_B \theta_N$, then the power within a frequency band of width W flowing past any point is $k_B \theta_N W$.

4.14. Use the union bound and equations (4.21) and (4.57) to show that the error probability of the PPM receiver scheme of Section 4.3.4 is bounded by

$$P_W < \exp(J'T_W - (1-\lambda)^2 PT_W/\eta) + \exp(-\lambda^2 PT_W/\eta)$$

when the threshold is set at $(1 - \lambda)\sqrt{(E/T_b)}$ $(0 < \lambda < 1)$. Hence show that reliable communication is possible in the limit $T_W \to \infty$ at any scaled rate J' less than P/η. (Hint: Choose λ so that the two terms on the right-hand side of the above inequality are equal.)

4.15. Derive (4.61) from (4.60).

5. Signal spaces

5.1 Expansions as function spaces

5.1.1 *Introductory*

We have already expressed signals as linear combinations of a fixed set of functions of time, with time-independent coefficients. Thus in Section 3.3.1 a pulse-train was written as

$$\Sigma_k a_k p_k(t), \qquad (5.1)$$

where p_k is a pulse in the kth interval from kT_b to $kT_b + T_b$. Earlier in Section 2.3.1 a periodic signal was expressed as a Fourier series

$$s(t) = \Sigma_{n=-\infty}^{\infty} S_n \exp(2\pi j n t/T). \qquad (5.2)$$

These expansions are analogous to the expansion of a vector in terms of a basis set of vectors and we represent the signal as a vector in a multidimensional space, the *signal space*. The expansion coefficients are coordinates specifying the vector in terms of the basis-set, the set of functions of t in terms of which the expansion is carried out. It is vital to realize that the signal-vector is specified by a set of time-independent coefficients and thus is a static quantity, not varying with time. It represents the entire signal over a specified interval.

The problem at the receiving end may be stated thus. The coefficients in the appropriate expansion of the transmitted signal must be determined, or in geometrical terms, the point in signal space used by the transmitter. One solution to this problem was described in Section 3.3.1. It was based on functions obeying the Nyquist property (3.8). A special method suitable for rectangular pulses was described in Section 4.3.2. A generalized version of these methods will be described shortly after a few preliminaries.

To keep the mathematics simple it is often convenient to assume that the signal is periodic with a period T very much greater than a typical time-scale (a pulse-length for instance) of the signal. It may be helpful to imagine the signal as an electromagnetic field circulating endlessly around an ideal waveguide or coaxial cable with a round-trip time of T (Section 2.3.5). Then appropriate measurements determine the state of the electromagnetic field, and hence the message 'written into' this field.

It should also be noted that the signals to be considered are usually the signals as they appear at the receiver. Thus when we talk about the energy of a signal we mean its energy at the receiver, not its energy at the transmitter, which may well be millions of times greater.

When discussing the theory in general terms it is better to permit the coefficients and basis sets to be complex-valued quantities, so that the algebra does not have to be rewritten for the manipulation of genuinely complex quantities. However real coefficients and real basis sets will be used in the study of geometric structures in signal space, like the signal-constellations of Section 3.4.4.

5.1.2 *Dimension and bandwidth equivalent*

The dimension of the signal-space will be defined informally as the number of real values needed to specify any signal adequately at the receiver. This is a finite number if the signal has a finite duration T. We consider some examples:

(a) For the real pulse-train (5.1) the number of real coefficients to be specified is T/T_b, where T_b is the pulse-duration.

(b) For the phasor pulse-train (3.24) the number of complex coefficients is equal to T/T_b, and hence the number of real values is $2T/T_b$.

(c) Let us consider a real base-band signal, limited to frequencies of magnitude less than f_M. The Fourier coefficients in (5.2) have to be specified for n from $-f_M T$ to $+f_M T$, so that (except for the zero-frequency term which is ignored) there are apparently $2f_M T$ complex values to be specified. However, since the signal is real the coefficients obey the reality condition $S_{-n} = S_n^*$ (2.46), so that they need only be specified for positive n. Hence only $f_M T$ complex values need to be specified or equivalently $2f_M T$ real values, so that the dimension is $2f_M T$.

(d) Finally let us consider a carrier-based signal, of period T, with bandwidth W. The Fourier coefficients in (5.2) need only be specified for positive n, because of the reality condition, and then only in a range WT, since outside this range they are zero. Hence the dimension is $2WT$.

In all these cases the dimensionality is proportional to T. This suggests we define a signalling rate as the constant of proportionality. It is more convenient however to define a signalling rate B as half the dimensionality divided by T. This rate will be called the *bandwidth equivalent*, for a reason that will be apparent in a moment, and it is the average rate at which complex quantities are to be measured at the receiving end. In the four examples given the bandwidth equivalents are respectively (a) $\frac{1}{2}T_b^{-1}$, (b) T_b^{-1}, (c) f_M, and (d) W. The factor half makes the bandwidth equivalent B the same as the bandwidth W for band-limited signalling and it gives a more natural definition when we consider the quantum theory of carrier-based signalling (Section 8.3.4). Conversely the dimension of the signal-space for a signal of duration T and bandwidth equivalent B is $2BT$.

The bandwidth equivalent is not the same as the information rate J in general. For pulsed carrier-based signalling as described in Section 3.4.4 the information rate is given by

$$J = B \log_2 Q \tag{5.3}$$

where Q is the number of points in the signal constellation.

5.1.3 Scalar product and length

In order to introduce the ideas of length and distance we need a 'scalar product'. For complex periodic functions $u(t)$, $v(t)$ of period T the scalar product (u, v) is most conveniently defined by

$$(u, v) = \int_{-T/2}^{T/2} u^*(t)v(t)\,dt. \tag{5.4}$$

For real functions the complex conjugation is not necessary, but there is no harm in keeping it, particularly as it provides a guide in certain manipulations. For localized pulses the limits may be set to $\pm\infty$. The squared 'length' of a signal u is taken as (u, u). This quantity is always real and non-negative, being zero only for the null-signal. The reason for putting in the complex-conjugation is to prevent a non-zero function from having a value of zero for its squared length. Some properties are (with c a constant):

$$(cu, v) = c^*(u, v),$$
$$(u, cv) = c(u, v),$$
$$(u^*, v^*) = (u, v)^*,$$
$$(u, v)^* = (v, u).$$

Implementing the computation of a scalar product is conceptually straightforward, involving a multiplier circuit to compute the integrand, followed by an 'integrate-and-dump' circuit. Since an integral like the one in (5.4) has the dimensions of energy when u and v are signals (see Section 2.1.3), the unit of length in signal-space is the square root of a joule.

Two functions u, v are called *orthogonal* if $(u, v) = 0$. A function satisfying $(u, u) = 1$ is said to be *normalized*. A set of functions $\{u_k\}$ is called orthogonal if $(u_k, u_l) = 0$ whenever $k \neq l$, and *orthonormal* if moreover $(u_k, u_k) = 1$ for all k. In this case we usually write

$$(u_k, u_l) = \delta_{kl} \tag{5.5}$$

where δ_{kl} is the Kronecker delta symbol defined by

$$\delta_{kl} = \begin{cases} 1 & \text{if } k = l, \\ 0 & \text{if } k \neq l. \end{cases} \tag{5.6}$$

Useful properties of the Kronecker delta are

$$\Sigma_l a_l \delta_{kl} = a_k \tag{5.7}$$

and

$$\Sigma_{kl} a_k b_l \delta_{kl} = \Sigma_k a_k b_k. \tag{5.8}$$

If a basis set is an orthonormal set, then it is called an orthonormal basis.

EXPANSIONS AS FUNCTION SPACES

Such a basis may be chosen in many ways, just as in ordinary vector theory. The basis functions are of course linear combinations of the defining basis set just like any function in the signal space.

The functions $\exp(2\pi j nt/T)$ from an orthogonal set of functions and the functions

$$e_n(t) \equiv T^{-1/2} \exp(2\pi j nt/T) \qquad (5.9)$$

form an orthonormal set. It should be noted that these functions provide a basis in which any reasonable set of periodic functions may be expanded, as a Fourier series in fact. Such a set of functions is called 'complete'.

Another set of orthonormal functions is provided by the rectangular pulses

$$p_k(t) = p(t - kT_b) \qquad (5.10a)$$

where

$$p(t) = \begin{cases} T_b^{-1/2} & \text{for } 0 < t < T_b, \\ 0 & \text{otherwise.} \end{cases} \qquad (5.10b)$$

Here T_b is T divided by some integer N, and the values of k range from 0 to $N-1$. Such a set is not complete, since only a limited class of functions can be expressed as a linear combination of these functions.

5.1.4 Expansions with an orthonormal basis

Let us suppose that a signal $s(t)$ may be expanded as

$$s(t) = \Sigma_k c_k u_k(t) \qquad (5.11)$$

in terms of a set $\{u_k\}$ of orthonormal functions. Then we may determine the coefficients c_k in this expansion by

$$c_k = (u_k, s). \qquad (5.12)$$

This is because

$$(u_k, s) = \Sigma_l (u_k, c_l u_l) = \Sigma_l c_l (u_k, u_l) = \Sigma_l c_l \delta_{kl} = c_k.$$

This method for determining the coefficients c_k is the same as the method described in Section 2.3.1 for determining the Fourier coefficients of a periodic function.

If the basis set u_k is not orthonormal it is possible to find another set of functions v_k in the signal space, i.e. linear combinations of the u_k, such that $(v_l, u_k) = \delta_{kl}$ (Exercise 5.2). Then by an analogous argument we obtain the coefficients in the expansion (5.11) as $c_k = (v_k, s)$. Thus although orthogonality is very convenient it is not absolutely necessary.

We thus have a conceptually simple signalling system. The signal $s(t)$ is set up as in (5.11) with the u_k prearranged, so that the coefficients c_k carry the information. Then the receiver has to compute the scalar products (5.12) for every value of k, to find the c_k from $s(t)$.

In the geometrical representation each signal $s(t)$ corresponds to some vector OP, where P is a point in signal space. Then we may regard the u_k as unit basis-vectors along a set of orthogonal axes, and then the coefficients c_k in the expansion

$$s = \Sigma_k c_k u_k$$

are the coordinates of P with respect to this basis. We also see that the square of the length OP is given both as (s, s) and as $\Sigma_k c_k^* c_k$, since

$$(s, s) = \Sigma_{kl} c_k^* \delta_{kl} c_l = \Sigma_k c_k^* c_k. \tag{5.13}$$

The square of the length is the energy of the signal. Thus if the energy is restricted to values less than E, the point P lies inside a hypersphere of radius $\sqrt{E}$.

5.1.5 Some examples of signal spaces

(a) Let us first consider a signal consisting of a single pulse of arbitrary height a_0

$$s(t) = a_0 p_0(t).$$

This formula is obtained from (5.1) by setting all the a_k to zero except a_0. The possible values of a_0 can be represented as points on a straight line. Hence the signal space is one-dimensional.

(b) A train of n rectangular pulses (5.10) forms the signal (5.1)

$$s(t) = \Sigma_{k=0}^{n-1} a_k p_k(t).$$

Since the pulses are non-overlapping they are automatically orthogonal. This signal may be represented as a point with coordinates $(a_0, a_1, \ldots, a_{n-1})$ in an n-dimensional space. Thus the signal constellations of Section 3.4.4 may be used to specify baseband signals; we set $n = 2$, pair off the pulses, and use the first and second pulse of each pair to send the x- and y-coordinates of the signal-point.

(c) In Section 3.4.4 we have considered a carrier-based pulse-train. Let us consider the signal represented by the phasor

$$\phi(t) = c_0 u(t),$$

a formula obtained from (3.24) by dropping all the terms except the one with $k = 0$. The actual signal is given by

$$s(t) = \sqrt{2} \{ \text{Re}(c_0) \text{Re}(z) - \text{Im}(c_0) \text{Im}(z) \}$$

with $z(t) = u(t) \exp(2\pi j f_c t)$. It will be shown in a moment that $\text{Re}(z)$ and $\text{Im}(z)$ are orthogonal functions of t. Hence if we use these functions to construct the basis set it is found that any signal can be represented as a point with coordinates $\text{Re}(c_0)$, $\text{Im}(c_0)$ in a two-dimensional space. Thus the signal constellations of Section 3.4.4 represent directly signal-points in a two-dimensional space.

EXPANSIONS AS FUNCTION SPACES

To show that Re(z) and Im(z) are orthogonal we first note that $(z, z^*) = 0$, because z contains only positive-frequency components, z^* only negative-frequency components, and because any positive-frequency component is orthogonal to any negative-frequency component (Section 3.4.2). Then substituting $z = \text{Re}(z) + j \text{Im}(z)$ gives the required answer.

5.1.6 Hadamard functions

An interesting set of orthogonal functions may be constructed out of rectangular pulses. These will be known as *Hadamard functions* after the Hadamard matrices from which they are constructed (MacWilliams and Sloane 1977, Section 2.3). These functions are a non-trivial example of an orthonormal basis and also an example of a binary error-correcting code in polar form.

Let the pulses have width T_b and let the word-length be $T_W = 2^m T_b$, where m is a positive integer, so that there are $M = 2^m$ pulses in each signal or 'codeword'. We set

$$u_k = M^{-1/2} \sum_{l=0}^{M-1} H_{kl} p_l(t), \quad k = 0, 1, \ldots, M-1, \qquad (5.14)$$

where the p_l are the orthonormal rectangular pulses given by (5.10), and where the coefficients H_{kl} have the values ± 1 so that each signal is a string of M polar pulses. These coefficients are defined as follows:

We start with the matrix

$$\begin{bmatrix} + & + \\ + & - \end{bmatrix}$$

where $\pm$ is short for ± 1, and copy it four times to form a square of twice the size. The bottom right-hand copy in this square is a negative. Then the process is repeated until a square matrix of size 2^m is built up. (Such matrices are examples of Hadamard matrices.) The first two iterations give:

$$\begin{bmatrix} + & + & + & + \\ + & - & + & - \\ + & + & - & - \\ + & - & - & + \end{bmatrix} \quad \begin{bmatrix} + & + & + & + & + & + & + & + \\ + & - & + & - & + & - & + & - \\ + & + & - & - & + & + & - & - \\ + & - & - & + & + & - & - & + \\ + & + & + & + & - & - & - & - \\ + & - & + & - & - & + & - & + \\ + & + & - & - & - & - & + & + \\ + & - & - & + & - & + & + & - \end{bmatrix}$$

If the rows and columns are labelled 0 to $M-1$ then this gives H_{kl}.

It is easy to show that the rows are orthogonal, i.e. that the 'dot product' of two rows $\Sigma_i H_{ki} H_{li}$ is zero if $k \neq l$. This is done by induction, that is, we prove that if the result is true for the size 2^m then it is true for the next size up as well. This chain is started at $m = 1$, where the result is obvious. Then suppose that the rows of the mth case, a $2^m \times 2^m$ array, are orthogonal. We consider the dot product of different rows of the $(m + 1)$th case, which are twice as long. If both rows are from the same half of the matrix, say the top half, then the second part of the dot product repeats the first part, which is a dot product of two rows of the mth case. Thus the answer is zero. If the rows are from opposite halves, then the second part of the dot product cancels out the first part, so that again the answer is zero.

As a consequence of this fact and of the orthonormality of the $p_l(t)$ in (5.10) it is readily shown that the $u_k(t)$ are also an orthonormal set (Exercise 5.7).

There is a short cut in evaluating the dot products $r_k = (u_k, r)$ with the received signal r. From (5.14) we find

$$r_k = (u_k, r) = \Sigma_l H_{kl} q_l, \qquad k = 0, 1, \ldots, (M-1) \tag{5.15a}$$

with

$$q_l = M^{-1/2}(p_l, r), \tag{5.15b}$$

quantities immediately available by 'integrate-and-dump'. We have to evaluate M scalar products and thus it seems that we have to perform M^2 additions or subtractions to work out the sums in (5.15a) over M values of l, for M values of k. Fortunately this is not necessary. It is readily shown that if k is written as a binary number $(k_{m-1}, k_{m-2}, \ldots, k_0)$ with

$$k = 2^{m-1} k_{m-1} + 2^{m-2} k_{m-2} + \ldots + 2k_1 + k_0$$

and similarly for l, then

$$H_{kl} = (-1)^{k_0 l_0} (-1)^{k_1 l_1} \ldots \tag{5.16}$$

If we write $r(k_{m-1}, k_{m-2}, \ldots, k_0)$ for r_k and similarly $q(l_{m-1}, l_{m-2}, \ldots, l_0)$ for q_l, then

$$r(k_{m-1}, k_{m-2}, \ldots, k_0) =$$
$$\Sigma_{l_0=0}^1 (-1)^{k_0 l_0} \Sigma_{l_1=0}^1 (-1)^{k_1 l_1} \ldots \Sigma_{l_{m-1}=0}^1 (-1)^{k_{m-1} l_{m-1}} q(l_{m-1}, \ldots, l_0).$$

The inside sum has to be done 2^m times, for each value of the bits k_{m-1} and $l_{m-2}, l_{m-3}, \ldots, l_0$. Each sum involves just one addition or subtraction. The next sum has to be done over l_{m-2} for each value of the bits k_{m-1}, k_{m-2} and $l_{m-3}, \ldots, l_0$. Thus again there are 2^m additions or subtractions in all. So it goes on for m steps altogether. Overall there are $m \cdot 2^m$ arithmetic operations, a number considerably less than the $2^m \cdot 2^m$ operations if the sums were done directly. (This algorithm is akin to the Cooley–Tukey algorithm for

EXPANSIONS AS FUNCTION SPACES

computing discrete Fourier transforms. See Oppenheim and Schafer 1979, Chapter 6.)

Example: With $M = 8$ let $(q_0, \ldots, q_7) = (0.6, -0.5, 0.4, -0.1, 0.2, 0.8, 0.3, 0.7)$. We lay the calculation out as follows, with the q_k in the left-hand column:

```
0.6  ╲  0.8  ╲  1.5  ╲  2.4
-0.5  ╳  0.3  ╳  0.9 ───  0.6
 0.4  ╳  0.7 ───  0.1     -0.2
-0.1  ╱  0.6    -0.3       0.4
 0.2 ───  0.4    0.5      -1.6
 0.8    -1.3    -2.1       2.6
 0.3     0.1     0.3      -0.2
 0.7    -0.8    -0.5       0.8
```

The lines show a typical sum-and-difference to be evaluated at each step. Thus $0.8 = 0.6 + 0.2$ and $0.4 = 0.6 - 0.2$. The final column gives the values r_0 to r_7 in order.

Overall we see that the computation of the set of scalar products (u_k, r) has been made into a two-stage process. The first stage (5.15b) is the evaluation of the scalar products with respect to a convenient basis, the pulse functions, carried out by 'integrate and dump'. Thereafter these values are processed as just described to evaluate the dot products in (5.15a). A two-stage evaluation is often much easier to implement than the conceptually simpler direct calculation.

5.1.7 Matched filters

Scalar products like

$$c_k = \int_{-T/2}^{T/2} u_k^*(t) s(t) \, dt \tag{5.17}$$

can often be computed by a process of filtering followed by sampling. The filter is called a *matched filter*, matched to the function $u_k(t)$. (It is not necessary that the $u_k(t)$ form an orthogonal set.) This is particularly convenient if the functions $u_k(t)$ are displacements of a single function $u_k(t) = u(t - kT_b)$ as in Section 3.3.1, because then only one filter is needed followed by sampling every T_b seconds. Each function $u_k(t)$ must be non-zero only in some interval τ'_k to τ''_k, say, lying within the period $-T/2$ to $T/2$. Thus the limits in (5.17) may be replaced by $-\infty$ to $+\infty$. We set up a filter with the impulse-response function

$$h_k(t) = u_k^*(\tau''_k - t). \tag{5.18}$$

so that

$$u_k^*(t) = h_k(\tau''_k - t) \tag{5.19}$$

We note that $h_k(t)$ vanishes for $t < 0$ and so obeys the causality condition (2.27). Then it follows that

$$c_k = \int_{-\infty}^{\infty} s(t) h_k(\tau_k'' - t) \, dt. \tag{5.20}$$

The right-hand side is just the convolution $s * h_k$ evaluated at τ_k'' (Section 2.1.3). Suppose now that the functions $u_k(t)$ have the form $u(t - kT_b)$ and that $u(t) = 0$ for $t > \tau_0''$. Then we find that $u_k(t) = 0$ for $t > \tau_k''$ where $\tau_k'' = \tau_0'' + kT_b$. Moreover, we find

$$h_k(t) = u_k^*(\tau_k'' - t) = u^*(\tau_0'' + kT_b - t - kT_b) = u^*(\tau_0'' - t). \tag{5.21}$$

Since this function does not depend on k it can be implemented by a single filter and the outputs are obtained by sampling at the moments $\tau_k'' = \tau_0'' + kT_b$. In particular it should be noted that the process of 'integrate and dump' implements matched filtering when the basis is a set of rectangular functions (Section 4.3.2).

5.1.8 The effect of additive white noise

Suppose that a signal of the form

$$s(t) = \Sigma_k c_k u_k(t)$$

is transmitted (with the c_k and the u_k real), and that before it is received white noise (with mean zero) is added so that an example of the received signal is $s(t) + x_\alpha(t)$ where $x_\alpha(t)$ is an example of noise (Section 4.2.1). Then the outputs formed by taking scalar products as in Section 5.1.4 are

$$c_{k\alpha}' = (u_k, s + x_\alpha) = c_k + N_{k\alpha}$$

where

$$N_{k\alpha} = (u_k, x_\alpha). \tag{5.22}$$

We wish to know the variances and covariances $\langle N_k^* N_l \rangle$ of the noisy outputs. (As usual we insert a complex conjugation as a guide, although in fact these quantities are real.) A simple result is obtained in the case when the noise is white.

We expand the basis set in terms of the functions (5.9) as

$$u_l(t) = \Sigma_{n=-\infty}^{\infty} U_{ln} e_n(t),$$

using the completeness property of these functions. Using the completeness property again we expand

$$x_\alpha(t) = \Sigma_n Y_{\alpha n} e_n(t).$$

If this is compared with the Fourier series (4.23) for x_α it is found by (5.9) that $Y_{\alpha n} = T^{1/2} X_{\alpha n}$. Hence (4.38) gives

$$\langle Y_m^* Y_n \rangle = \tfrac{1}{2}\eta \delta_{mn}$$

where η is the one-sided spectral density of the white noise. Next we expand the scalar product (5.22)

$$N_{k\alpha} = (u_k, x_\alpha) = \Sigma_{mn} U_{km}^* (e_m, e_n) Y_{\alpha n}$$
$$= \Sigma_{mn} U_{km}^* \delta_{mn} Y_{\alpha n} = \Sigma_n U_{kn}^* Y_{\alpha n},$$

the sums over m and n being taken from $-\infty$ to $+\infty$. Thus we find

$$\langle N_k^* N_l \rangle = \Sigma_{mn} U_{km} U_{ln}^* \langle Y_m^* Y_n \rangle = \Sigma_{mn} U_{km} U_{ln}^* \cdot \tfrac{1}{2}\eta \delta_{mn}$$
$$= \tfrac{1}{2}\eta \Sigma_n U_{kn} U_{ln}^* = \tfrac{1}{2}\eta (u_l, u_k).$$

At the last stage we have used the result

$$(u_l, u_k) = \Sigma_n U_{ln}^* (e_n, e_m) U_{km} = \Sigma_n U_{lm}^* \delta_{nm} U_{km} = \Sigma_n U_{ln}^* U_{kn}.$$

Thus finally by the orthonormality of the u_k we find

$$\langle N_k N_l \rangle = \tfrac{1}{2}\eta \delta_{kl}. \qquad (5.23)$$

Thus these random variables are uncorrelated and have the same variance $\tfrac{1}{2}\eta$. This generalizes a result from Section 4.3.2 in the special case when the basis functions were rectangular pulses, although at that stage a different normalization was employed. (For an alternative derivation of (5.23) see Exercise 5.11.)

A similar argument proves that the scalar product (v, x) of white noise x with any real normalized function v always has a variance $\tfrac{1}{2}\eta$. Thus to detect the presence or absence of some specified signal u in the presence of white noise it pays to choose v so that (v, u) is maximized (subject to $(v, v) = 1$). In geometrical terms we should choose v parallel to u, so that v is equal to u times a normalization constant.

Example: Binary signalling is often carried out by choosing between one of two signals $s_1(t)$ and $s_2(t)$ in any given interval, say $t = 0$ to T_b. These signals need not be orthogonal. We may suppose that $s_1(t)$ is always there, but when $s_2(t)$ is to be used the amplitude $s_2(t) - s_1(t)$ is added to it. Hence we wish to detect the presence or absence of $u = s_2 - s_1$ and so we use a filter matched to this function.

A very important consequence, simply expressed in geometric terms, is the following: The variance of any component of noise with spectral density η is $\tfrac{1}{2}\eta$. Hence by (4.17) the probability of Gaussian noise displacing a signal point by more than a distance x *in a specified direction* is $\tfrac{1}{2}\operatorname{erfc}(x/\sqrt{\eta})$.

5.2 Codes in signal-space

5.2.1 *Introductory*

The idea of representing possible signals as points in a signal-space may be further developed. Suppose that we wish to send one of L possible real signals

$s_p(t)$ ($p = 1, 2, \ldots, L$). These will be called 'codewords'. They are presumed to vanish unless $0 < t \leqslant T_W$, that is they all last for a word-time T_W. (Obviously the signalling system is not used just once but repeatedly every T_W seconds. If periodicity is required, then the periodic time T is chosen as a sufficiently large multiple of T_W.) It is assumed that the signals may be expanded using a real orthonormal set of functions $u_k(t)$ in an M-dimensional real space:

$$s_p(t) = \Sigma_{k=1}^{M} c_{kp} u_k(t) \qquad \text{for } p = 1, \ldots, L. \tag{5.24}$$

The functions u_k also vanish unless $0 < t \leqslant T_W$, so that the scalar product of such functions may be defined as in (5.4) with the limits replaced by 0 and T_W. Thus each signal s_p may be regarded as a vector joining the origin to the signal-point S_p with coordinates $(c_{1p}, c_{2p}, \ldots, c_{Mp})$. These signals are also subject to a constraint that the average power is limited so that it cannot exceed some value P. The total energy available for the interval T_W is thus less than or equal to PT_W, and hence we have

$$(s_p, s_p) \leqslant PT_W.$$

Thus we find (by (5.13))

$$\Sigma_k c_{kp} c_{kp} = \Sigma_k c_{kp}^* c_{kp} = (s_p, s_p) \leqslant PT_W. \tag{5.25}$$

Hence the signal-points lie inside or on an 'M-sphere' (a hypersphere in M dimensions) of radius $\sqrt{(PT_W)}$ centred on the origin. This sphere will be called the *energy sphere*. It seems reasonable to try and pack as many signal-points into this M-sphere as possible, but the minimum spacing has to be maintained at a value large enough to give reliable signalling in spite of the noise.

The operation of *decoding* is deciding from the received signal $r(t)$ which of the possible L signals $s_1, \ldots, s_L$ was actually used. The best strategy is called Maximum Likelihood Decoding (MLD), and it attempts to answer the question: 'Given the received signal, what is the most likely signal to have been sent?' However, provided that all the L signals are equally likely to have been used, then some elementary probability theory shows that we may answer the following question instead: 'Which of the possible transmitted signals s_p makes the signal that was received, namely $r(t)$, the most likely?' (See Exercise 5.12.)

We suppose that the signal has been sent over an AWGN channel, which adds noise of spectral density η. The received signal r will not in general be expressible as a linear combination of the u_k, since these are not necessarily a complete set. None the less we shall assume that there is a larger orthonormal basis $v_1, v_2, \ldots, v_N$ in terms of which any possible received signal including r and $u_1, \ldots, u_M$ may be expanded. (Some of these new vectors may be the same as the u_k.)

We determine the coordinates $r_k = (v_k, r)$ ($k = 1, \ldots, N$) of the received signal r corresponding to the point R in the space with the v_k as a basis set (Fig. 5.1). Suppose that s_p was the signal actually used. Then the coordinates

CODES IN SIGNAL-SPACE

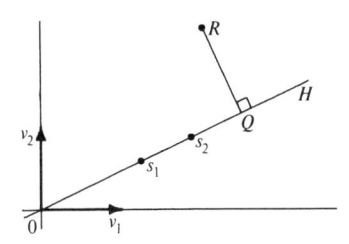

FIG. 5.1 Minimum-distance decoding: nearest signal-point s_p to the received point R is also nearest to Q in the hyperplane H in which all the signal-points lie.

of the noise in this instance are $x_k = (v_k, r - s_p)$. The components of the random noise-signal $X(t)$ with respect to the basis-set $v(t)$ are the random variables $X_k = (v_k, X)$ which are independent identically distributed Gaussian variables with variance $\frac{1}{2}\eta$ (as in (5.22) and (5.23)). Thus the probability density for finding $(X_1, \ldots, X_N)$ at $(x_1, \ldots, x_N)$ is $(\pi\eta)^{-N/2} \exp(-\sum_{k=1}^{N} x_k x_k/\eta)$. But by the energy theorem (5.13) the sum $\sum_k x_k x_k$ is $(r - s_p, r - s_p)$. So the choice of signal s_p which maximizes this probability density is the one which is nearest the received point. This method of choice is naturally called *minimum distance decoding*.

We would like a criterion in terms of distances in the sub-space with the u_k as basis. This space is a hyperplane H (Fig. 5.1) immersed in the big space, and it passes through the origin since the null-signal lies in this space. We simply drop a perpendicular RQ into this hyperplane so that Q lies in the hyperplane. Then the signal-point nearest R is also the signal-point nearest Q. This geometric picture may be used to guide the algebraic discussion. We choose

$$q(t) = \Sigma_l u_l(t)(u_l, r)$$

as a signal in the M-space, which in fact corresponds to Q. Then we find that

$$(u_k, r - q) = 0 \qquad (5.26)$$

for all k from 1 to M, since

$$(u_k, q) = \Sigma_l (u_k, u_l)(u_l, r) = \Sigma_l \delta_{kl}(u_l, r) = (u_k, r).$$

In geometric terms equation (5.26) implies that RQ is perpendicular to every basis vector in H, hence to any vector in H, and hence to H itself. Next we see that

$$s_p - r = (s_p - q) - (r - q)$$

and so by squaring we find

$$(s_p - r, s_p - r) = (s_p - q, s_p - q) + (r - q, r - q) - 2(s_p - q, r - q).$$

Since $s_p - q$ is a linear combination of the u_k the last term on the right-hand side vanishes by (5.26). So we simply have to find the s_p which makes the first term the smallest, since the second term is fixed. The first term is readily expressed in terms of scalar products of r and s_p with the u_k. (See Exercise 5.13.)

5.2.2 *Lattice codes*

The idea of the signal points being arranged in a regular lattice as in Section 3.4.4 (Figs 3.11(f) and 3.11(g) may be extended to more than two dimensions. Typically one would do this in a situation where the noise-level is rather low, or where the bandwidth or signalling rate is limited. Thus one tries to pack as many points as possible inside the energy-sphere whose radius is determined by the constraint on the average power, but at the same time keeping the minimum distance between them large. Thus a 'tight-packed' lattice is preferable. From this point of view it is better to use a hexagonal lattice like that in Fig. 3.11(g) rather than a square lattice, since the former gives a density of points greater by a factor $\sqrt{(4/3)}$ for the same minimum spacing. In a higher number of dimensions greater savings are possible. Thus in four dimensions we may take the hypercubic integer lattice Z_4, where any string of four integers specifies a lattice-point, and turn it into the tight-packed lattice D_4 by allowing only those points of Z_4 where the coordinates add up to an even number (Sloane 1981b). Let us suppose that the signal points fall on this lattice. The process of finding the nearest lattice point in Z_4 and hence the nearest signal-point to the received-signal point R is just a matter of rounding each coordinate of R to the nearest integer value, but in D_4 this may leave us with an odd value for the sum of the coordinates. In that case one of the coordinates is rounded the wrong way, the one which is furthest from an integral value and hence the least reliable. Thus if (1.57, 0.63, −0.29, 1.92) is received, simple rounding gives (2, 1, 0, 2) and these integers add up to the odd number 5, so we have to round one of the values the wrong way. The first coordinate is evidently the furthest from an integral value, and so it is rounded the wrong way to give (1, 1, 0, 2). It is easy to verify that this algorithm gives the lattice point nearest R (Exercise 5.14).

In eight dimensions there is a tight-packed lattice E_8 formed as follows: First the integer hypercubic lattice Z_8 is thinned out to D_8 by the condition that the eight integer coordinates add up to an even number. Then a shifted lattice D_8' is formed by displacing each point of D_8 by the vector

$$(\tfrac{1}{2}, \tfrac{1}{2}, \tfrac{1}{2}, \tfrac{1}{2}, \tfrac{1}{2}, \tfrac{1}{2}, \tfrac{1}{2}, \tfrac{1}{2}).$$

Then E_8 is the lattice formed by all the points in D_8 and D_8'. (No attempt will be made to prove that this construction produces a lattice where every point is equivalent to every other, let alone that it produces a tight-packed lattice. For further details the reader is referred to articles by Sloane 1981, 1984.)

To find the lattice point nearest the received point R we find the nearest point P in D_8 by the same procedure as was described for D_4, and then the nearest point P' in D_8'. This is done by subtracting $\tfrac{1}{2}$ from each coordinate of R, process the result as for D_8, and then add $\frac{1}{2}$ back to each coordinate. The nearer of P and P' to R is the answer.

We note four points in passing:

(a) The decoding algorithm is incomplete as it does not prescribe what to do if the nearest point is outside the energy-sphere and so cannot correspond to any signal sent by the transmitter.

(b) The number of points inside the energy-sphere is not generally a power of 2, and hence the method is a little clumsy as a way of encoding bit-strings.

(c) There is a remarkable tight-packed lattice in 24 dimensions known as the Leech lattice, of great interest to mathematicians.

(d) Choosing the nearest lattice point to an arbitrary point in space may be used for analogue-to-digital conversion of multi-component vector analogue signals. Such a process is thus called 'vector quantization' and is a generalization to several dimensions of the method of quantization described in Section 3.2.6 (Gersho and Cuperman 1983).

5.2.3 *Spherical codes*

In a *spherical code* the signal points lie on the energy-sphere of radius $\sqrt{(PT_W)}$, not inside it. A modified decoding system known as correlation decoding may then be used. As usual we look for the signal point s_p which minimizes the distance from the received point r. Thus we minimize

$$(s_p - r, s_p - r) = (s_p, s_p) - 2(s_p, r) + (r, r).$$

The first and third terms on the right-hand side are independent of s_p, the first always being equal to PT_W. Thus we choose the s_p which maximizes the scalar product (s_p, r). Such scalar products are sometimes called 'correlations', and hence the name for this method of decoding. It should be noted that the s_p need not form an orthogonal set. Many of the modulation and coding schemes in this book give rise to spherical codes.

5.2.4 *Orthogonal and biorthogonal signalling*

In Section 4.3.4 we examined a method of signalling by Pulse Position Modulation (PPM) which in principle allows us to send information at a finite rate with a reliability as high as we please. By using the theory of signal spaces we may fit this method into a more general scheme known as *orthogonal signalling*. It will be assumed that the noise is additive white Gaussian noise with spectral density η.

With a bandwidth equivalent B over a time T_W we may represent a signal as a point in a real space of $M = 2BT_W$ dimensions. We select an orthonormal set of real functions u_k ($k = 1, \ldots, M$) and choose the transmitted signal s as $\pm cu_k$ for a particular value of k and a particular choice of sign. Since the average power P is predetermined it follows that $c = \sqrt{(PT_W)}$, since (s, s) is the total energy PT_W. The information conveyed is then $\log_2(2M)$ bits, and the information rate J (in bits per second) is given by

$$J' = J \ln 2 = T_W^{-1} \ln(2M). \tag{5.27}$$

If the signal-set consists only of the positively signed orthogonal functions cu_k the scheme is known as orthogonal signalling. If both signs are used as just described then the scheme is known as biorthogonal signalling.

The PPM system of Section 4.3.4 is a particular example of an orthogonal scheme, since the basis-set consists of the orthonormal pulse-functions p_k (5.10). The bandwidth equivalent (Section 5.1.2) is related to the pulse-duration by $B = 1/(2T_b)$. The scheme becomes biorthogonal signalling if the pulses are permitted to have either sign. In another scheme the Hadamard functions of Section 5.1.6 are used for the basis-set. Every signal is a polar pulse-train and is thus a binary code in polar form. This system has the advantage over the PPM system that the power is constant, and the energy is not released in huge bursts. Thus the first objection to PPM signalling (Section 4.3.4) is easily overcome by the use of a different basis set. The disadvantage is that the decoding operation is more complex.

Suppose then that the signal $s(t) = +\sqrt{(PT_W)} \cdot u_1(t)$ is sent. It corresponds to the point $S_1 = (\sqrt{(PT_W)}, 0, 0, \ldots)$ in signal-space (Fig. 5.2). The received signal $r(t)$ corresponds to a point R, which as described in Section 5.1.8 is displaced from S_1 by a vector whose components form a set of independent identically

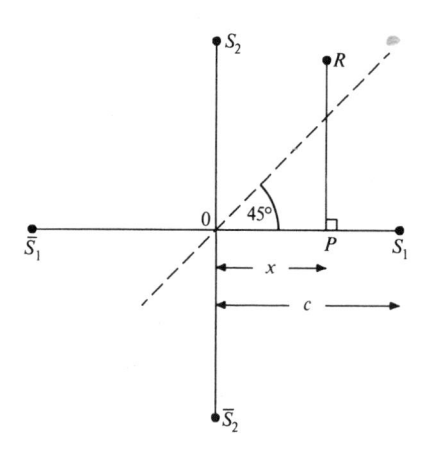

FIG. 5.2 The tangential-union bound for biorthogonal codes. Here S_1 is the point corresponding to the signal that was sent. An error occurs if the noise causes the received point R to be displaced perpendicular to OS_1 by a distance greater than OP.

distributed Gaussian random variables each with variance $\frac{1}{2}\eta$. The minimum-distance-decoding decision strategy is to pick that signal point S_p or $\bar{S}_p$ ($p = 1, \ldots, M$) nearest R. (Here S_p and $\bar{S}_p$ are respectively the points displaced positively and negatively from the origin by $\sqrt{(PT_W)}$ along the pth axis.) Since the code is a spherical code this may be implemented by finding which scalar product (u_k, r) is the largest in magnitude (Section 5.2.3). (See Exercise 5.20.)

We now consider an upper bound on the error-probability P_W, in order to see under what conditions it may be made arbitrarily small. (The subscript W denotes that this is a probability of a decoding error for the code-word, rather than for a single bit.) Actually an exact formula may be found in this instance (Appendix A) but a good approximate method, Berlekamp's tangential-union bound (1980), will be described, which can be used in a wider context. It is a modification of the union bound which gives tighter results especially when the noise-level and hence the error-rate are high. The ordinary union bound can give too high an estimate for the error-probability, and in consequence is pessimistic about the maximum information rate of orthogonal signalling. (Compare the results of Exercise 5.16 and Exercise 4.14.)

An error occurs if as in Fig. 5.2 the point R is nearer some other point (S_2 in this case) than it is to the correct point S_1. If R is nearer S_2 than S_1 it is on the 'wrong' side of the hyperplane which bisects S_1S_2 perpendicularly, and which is shown as a dotted line in Fig. 5.2. The noise-vector S_1R is broken into the sum of two parts, the radial component S_1P and a 'tangential' component PR. These components are independent random variables. This is because noise-components along orthogonal directions in signal-space are uncorrelated (5.23) and because uncorrelated mutually Gaussian random variables are independent. Suppose for the moment that OP is fixed at some value x. Then an error occurs if PR exceeds the value x in the direction OS_2, or indeed in any one of $2M - 2$ directions parallel to or antiparallel to any axis except OS_1. So by the union bound the probability $P_e(x)$ of this happening is less than or equal to

$$F(x) = \min\{1, 2M \cdot \tfrac{1}{2} \operatorname{erfc}(x/\sqrt{\eta})\}, \tag{5.28}$$

where the function $\min(x, y)$ is defined as x if $x \leqslant y$ and y if $y < x$. We have used the result at the end of Section 5.1.8. We have also replaced $2M - 2$ by $2M$ and used the fact that a probability cannot exceed the value 1.

The probability of finding OP in the range x to $x + dx$ is

$$\rho(x)\,dx = (\pi\eta)^{-1/2} \exp\{(x-c)^2/\eta\}dx. \tag{5.29}$$

where

$$c = \sqrt{(PT_W)} = \sqrt{\{(P/J')\ln(2M)\}} \tag{5.30}$$

by (5.27). Thus the probability that OP is in the range x to $x + dx$ *and* that there is an error is the product of $\rho(x)\,dx$ and $P_e(x)$ and hence less than or equal to $\rho(x)F(x)\,dx$. Thus integrating over all possible values of x bounds the overall word-error probability by

$$P_W \leq \int_{-\infty}^{\infty} \rho(x)F(x)\,dx. \qquad (5.31)$$

The factors F and ρ are shown in Fig. 5.3. At some value x_0 the function F suddenly drops away from unity. If we use the approximation (4.21)

$$\tfrac{1}{2}\,\text{erfc}\,(x/\sqrt{\eta}) < \exp(-x^2/\eta)$$

(valid for $x > 0$) we see that

$$2M\exp(-x_0^2/\eta) > 1$$

and hence that

$$x_0 < \sqrt{\{\eta \ln(2M)\}}$$

with the ratio of the two sides approaching unity as M becomes very large.

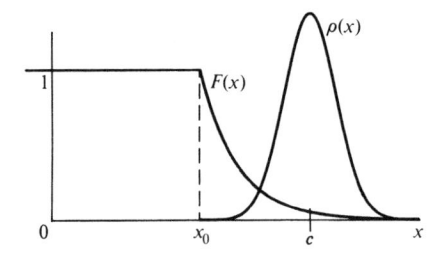

FIG. 5.3 The two factors $F(x)$ and $\rho(x)$ in the integral for the tangential-union bound.

The area under $\rho(x)$ is unity, and it is apparent that for $c < x_0$ (that is for $P/J' < \eta$) the bound is close to unity and so is not useful. On the other hand for $c > x_0$ it is evident that c moves further and further to the right of x_0 as M becomes very large. Thus the error-probability falls to very low values. In this way it is readily shown that signalling with an arbitrarily small probability of error is possible if $P/J' > \eta$, i.e. if $J' < P/\eta$. The exact formula (Appendix A) can be used to shown that J' has to be less than P/η for reliable signalling.

There are several points worth noting:

(a) The choice of orthonormal basis is arbitrary provided the noise is white. Typical choices are the functions e_n (5.9) and the rectangular pulses p_k (5.10). In the latter case the coding is pulse-position modulation but with a choice of sign. For large values of M the pulse-power becomes enormous, and there are problems associated with limitations on the peak-power. However other orthogonal functions do not suffer from this problem.

(b) The bandwidth equivalent B, which determines the bandwidth, is equal to $M/(2T_W) = (4T_W)^{-1} \exp(JT_W)$. This grows exponentially with T_W, if J is kept fixed.

(c) The receiver has to calculate M scalar products and has to compare them at least. Since $M = \frac{1}{2}\exp(JT_W)$ the amount of calculation (the computing complexity) also grows exponentially with T_W.

(d) It may be shown from (5.30) and (5.31) that with P, J, and η fixed the error probability falls exponentially with T_W increasing, as

$$P_W < \exp(-\alpha T_W) \tag{5.32}$$

for some value of α expressible in terms of P, J, and η. A similar formula (4.58) has already been discussed in Section 4.3.4 in connection with reliable signalling, and the same conclusion applies, that the error probability falls off as a small negative power of M, which is a measure of the computing complexity. For if we use (5.27) to substitute for T_W in (5.32) we find

$$P_W = (2M)^{-\alpha/J'}, \tag{5.33}$$

much as we previously obtained (4.59). Thus orthogonal signalling is not a good method for achieving very high reliabilities, with specified values of P, η, and J' (with $J' < P/\eta$).

5.3 Shannon's theorem

5.3.1 *Introductory*

Shannon's work, published in 1948, started the whole field of information theory and by showing that reliable communication against a noisy background at finite rate was possible issued a challenge to coding theorists. It will be sufficient in the present context to describe Shannon's theorem as it applies to the channel with additive white Gaussian noise and to outline its derivation.

It has already been shown in Section 5.2.4 that reliable communication is possible in principle by the use of orthogonal functions. A much more general signalling system is now discussed.

5.3.2 *The theorem*

We imagine that we have an AWGN channel specified for us, with the bandwidth or bandwidth equivalent B laid down in advance, together with a limit P on the average power and a value η for the white-noise spectral density. We are permitted to choose our code, the period T_W for which every code-word is to last, and the information rate J (in bits per second). The error-probability P_W is to be reduced as far as possible. By the definition of the bandwidth equivalent the signal space has dimensionality.

$$M = 2BT_W \tag{5.34}$$

112   SIGNAL SPACES

and we have to choose

$$L = \exp(J'T_W) \tag{5.35}$$

points in this space, lying inside the energy M-sphere of radius $\sqrt{S}$ centred on the origin. Here $J' = J\ln 2$ is the scaled information rate (4.56), and

$$S = PT_W \tag{5.36}$$

is the total energy permitted for each code-word or signal. We expect that M will be very large, as large values of T_W may be used.

A code is set up by locating signal points at random inside the M-sphere SS of radius $\sqrt{S}$ centred on the origin O (Fig. 5.4). Then any particular signal-point A almost certainly lies just inside the surface of the M-sphere. This is

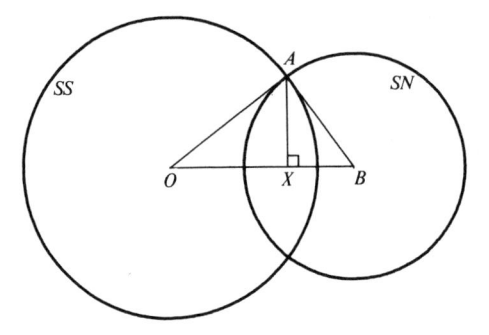

FIG. 5.4 Shannon's theorem: the signal A is sent and B is received. With very high probability OA is close to $S^{1/2}$, AB to $N^{1/2}$, and OB to $(S + N)^{1/2}$:

because most of the volume of the M-sphere lies very close to the surface when M is very large. This may be seen as follows: the ratio of the volume of an M-sphere of radius $0.999\sqrt{S}$ to the volume of SS is $(0.999)^M$. For $M = 10\,000$ this is 45×10^{-6} and for $M = 15\,000$ it is 0.3×10^{-6}. Thus for values of M in this range all but a few parts per million of the volume SS lies outside 99.9 per cent of its radius. In this outline the point A will simply be taken as lying on the surface of SS.

When the receiver measures the coordinates of the received signal there will be a variance $\frac{1}{2}\eta$ in each one because of the noise, which may be interpreted as a noise-energy. Thus the total noise-energy is

$$N = M \cdot \tfrac{1}{2}\eta = BT_W\eta. \tag{5.37}$$

The measured point B will almost certainly be found displaced from the true point A by a distance close to $\sqrt{N}$. Again for larger values of M the statistical uncertainties will be ignored and this distance will be taken as $\sqrt{N}$. The distance OB is determined by the total energy, signal S plus noise N, and so it will be taken as $\sqrt{(S + N)}$, again with the statistical fluctuations ignored. Thus by

Pythagoras's theorem the angle OAB is a right-angle (Fig. 5.4). (It might seem that B is heavily constrained, but actually it is free to choose any position on a 'ring' with OA as axis, which gives it freedom to move in $(M-2)$ dimensions.)

The decoder uses minimum-distance decoding and so it looks for the signal-point nearest B (Section 5.2.1). If this signal point is some point other than A then a decoding error occurs. We wish to know the probability of this happening. Let us consider the chances that a given signalling point A' (not the same as A) located at random inside the sphere SS is also found inside the sphere SN, the sphere of radius BA centred on B. (This radius is $\sqrt{N}$ if statistical fluctuations are ignored.) The probability is just the ratio of the lozenge-shaped volume in common to both spheres SS and SN to the volume of SS. The lozenge-shaped volume is less than the volume of a sphere of radius XA centred on X and the ratio of the volume of this sphere to that of SS is $(XA/OA)^M$. The ratio XA/OA is $\sqrt{\{N/(S+N)\}}$, as may be seen as follows: OAB is a right-angled triangle with sides $OA = \sqrt{S}$, $AB = \sqrt{N}$, and $OB = \sqrt{(S+N)}$, and its area is $\frac{1}{2}OA \cdot AB$. But its area is also $\frac{1}{2}XA \cdot OB$ so that $XA/OA = BA/OB$. Thus the probability is bounded above by $\{N/(S+N)\}^{M/2}$. Now with $L-1$ signal points other than A scattered around inside SS the union bound shows that the error probability P_W is less than $L\{N/(S+N)\}^{M/2}$. The logarithm of P_W satisfies

$$\ln P_W < \ln L - BT_W \ln(1 + S/N) = T_W\{J' - B\ln(1 + S/N)\}, \quad (5.38)$$

by (5.34) and (5.35). Thus provided that

$$J' < C' \quad (5.39)$$

where

$$C' = B\ln(1 + S/N), \quad (5.40)$$

we may make $\ln P_W$ as negative as we please by making T_W large enough. There is of course a considerable amount of tightening up needed to turn this outline into a proper proof. The interested reader is referred to Wozencraft and Jacobs 1965, Section 5.5.

This inequality may also be written as

$$J < C \quad (5.41)$$

with

$$C = B\log_2(1 + S/N) \quad (5.42)$$

where C is called the channel capacity in bits per second. We note that

$$C' = C\ln 2 \quad (5.43)$$

is the channel capacity scaled.

Shannon proved that signalling with an arbitrarily high reliability is possible at any rate less than C. Although (5.42) resembles formulae for 'moderate reliability' as discussed in Section 3.2,2, there was no hint at that stage that arbitrarily high reliability was possible, without J going to zero.

5.3.3 Random coding

It is worth examining a little more carefully what is meant by the signal points (or codewords) being located 'at random'. A more precise description of the theory involves the averaging of the error probability over all possible locations of the L signal points; that is, over all possible distributions of L points inside the sphere SS. Thus if the averaged error can be made as small as we please, then there must be at least one arrangement of the L points which can do better.

5.3.4 The converse theorem

Another question is whether it is possible to signal reliably at a rate higher than the maximum obeying the inequality (5.41). After all, the treatment of the inequalities has been fairly rough and the right-hand side of (5.41) as given by (5.42) may be too low. However, it can be shown that if the noise is additive white Gaussian noise, then it is not possible to violate (5.41) and still have reliable communication. It may be possible to do better in the presence of other types of noise. To take an extreme example: suppose the noise comes in huge bursts whose times are known in advance, with no noise in between. Then a very high rate of information may be obtained by communicating only during the noise-free periods. This is so even if the averaged noise-power is very high.

5.3.5 Reliability and complexity

We now consider how the probability of error and the computing complexity vary with the number of symbols $M = 2BT_W$ in each codeword (as T_W is varied, keeping B, J, and η fixed). The negative exponential dependence of P_W on T_W in (5.38) and on M through (5.34) survives a rigorous treatment and it is found that

$$P_W \sim \exp(-GM) \tag{5.44}$$

where G is the 'reliability exponent' (Gallager 1968). As the information rate J approaches the capacity C from below, the reliability exponent becomes very small.

Since Shannon's theorem is true for the average over all possible codes it would be surprising if nothing as good as the average could be found after a few attempts. None the less the encoding and decoding complexity would be huge. At the decoding end every one of L codewords would have to be checked against the received word, to find the nearest. Unfortunately, just as the error probability falls exponentially with the codeword time T_W, so also the value of L increases exponentially with T_W. Thus the error probability goes, roughly speaking, as a small negative power of P_W, as it did in Section 4.3.4. and in Section 5.2.4 in connection with orthogonal signalling. Thus high reliabilities are not practical with such a decoding scheme.

5.3.6 Wide-band and narrow-band signalling

By (5.40) the scaled capacity C' ($=C \ln 2$) increases steadily towards P/η as the bandwidth or bandwidth equivalent B increases without limit. It achieves 81 per cent of this value when B is twice P/η. Thus values of B around a few times P/η should be used, other things being equal. There is however no fundamental reason for increasing B far beyond this point, which unfortunately is what orthogonal signalling has to do.

If B is severely reduced below P/η, then the capacity falls, usually to a few times B, because in practice it is very hard to make logarithmic factors exceed about ten. As a consolation we may talk about a low-noise channel since then $B\eta \ll P$, but if we are permitted to increase the bandwidth, then this is preferable to using the multilevel techniques or lattice-codes previously described. Narrow bandwidths may be forced on us by protocols, legal constraints, or simply the speed of our electronics.

Signalling with a bandwidth equivalent very much greater than the information rate is known as *spread-spectrum* signalling, and it has many applications. Two military applications are the mitigation of the effects of jamming by an enemy, and the concealment of the existence of a transmitter or transmission. These are now considered very briefly:

(a) If the noise N as well as the signal-power S is independent of the bandwidth B, then the channel capacity C is proportional to the bandwidth, according to (5.42). This is typical of what happens if an enemy tries to jam a transmission by producing what is to all intents and purposes noise. He has to cover the bandwidth of the transmission but his total power is limited. Then the wider the bandwidth used the better. One method of doing this is to use pseudorandom polar signals akin to the Hadamard functions of Section 5.1.6. The enemy must not know the details of the coding scheme for otherwise he could interfere with the transmission very effectively by imitating it.

(b) Spread-spectrum signals look like white noise (limited of course to the transmission bandwidth) to a receiver that does not have the details of the coding scheme. Hence they are camouflaged by the ordinary background noise produced thermally and by other means.

In a third more peaceful use of spread-spectrum signalling each of several senders occupies the whole of a common frequency band, but he uses a coding scheme unique to him. A receiver can select a sender by matching his decoder to that sender's code. Since the spread-spectrum signals do not form an orthogonal set, but appear more like points chosen at random in a signal-space each receiver is interfered with by all the other senders. However this interference appears like noise, and reliable communication is possible provided that there are not too many senders using the spread-spectrum channel at once. This way of sharing a common channel is known as Code Division Multiplexing (CDM).

For further details see Cook and Marsh (1983), Dixon (1976), Holmes (1982), and the collection of articles in *IEEE* (1982*b*).

5.3.7 *To code or not to code*

Before any specific coding schemes are discussed, it is worth considering what are the advantages and disadvantages of coding, at least on an AWGN channel. This can be done because Shannon's theorem specifies what is the best that can be hoped for from coding without the need for considering particular codes.

The basic advantage of coding on the AWGN channel is that a given level of reliability can be achieved at a lower power level than without coding. The higher the degree of reliability required the more coding can offer. One way of reducing the effects of noise is to increase the received power. However, the cost of bigger power-amplifiers and better antennae may be greater than the cost of digital hardware for encoding and decoding, and it seems likely that the cost of the latter sort of electronics will keep falling faster than the cost of the former sort, so that error-correcting codes are bound to grow more popular.

There are disadvantages with using error-correcting codes:

(a) A very powerful code is easily upset by a slight degradation in the channel. Thus a slight loss of signal for any reason may cause an increase of several orders of magnitude in the error probability. This is most obviously seen if we imagine that we are using a very powerful code of the sort promised by Shannon. For an information rate just below capacity the error-probabilities are very low indeed. But if the information rate goes slightly above capacity, then the error-probability goes up to near unity (according to a theorem converse to Shannon's theorem). This would happen, for instance, if the noise-level were to increase slightly. This situation will be illustrated in due course in the next chapter (Fig. 6.5).

(b) Complicated encoding and decoding procedures may lead to comparatively long delays. This may be a disadvantage if the communication link is used in a control loop or in some similar real-time application. It would scarcely matter in a satellite link where there are long delays inevitable because of the finite speed of light.

(c) The throughput of the channel is limited by the speed at which the decoder operates. If high transmission rates are required, then it may not be possible to use a code which needs a lot of computation at the decoding end.

To see what savings in power are theoretically possible we examine two extreme cases, the first where the power available is the main constraint, with the bandwidth or bandwidth equivalent a secondary consideration. (This was a situation typical of satellite links in the early days. With the proliferation of such links this happy disregard of constraints on the bandwidth has not lasted.) The second situation is the obverse case, where the bandwidth is strictly limited and the power a secondary consideration.

(a) *The power-limited situation.* It is convenient to introduce two parameters. The first is γ_b, the energy needed for each bit of information, divided by the noise spectral density:

$$\gamma_b = P/(\eta J). \tag{5.45}$$

The second parameter κ might be called the bandwidth compression. It is the information rate divided by the bandwidth equivalent:

$$\kappa = J/B.$$

Shannon's bound (5.41) and (5.42) may then be written as

$$\kappa < \log_2(1 + \gamma_b/\kappa)$$

so that

$$\gamma_b > \gamma_S \tag{5.46}$$

where

$$\gamma_S = (2^\kappa - 1)/\kappa. \tag{5.47}$$

A small value of κ denotes a large bandwidth. It is readily shown from (5.46) and (5.47) that no matter how small κ is, i.e. no matter how large the bandwidth, the value of γ_b must exceed $\ln 2 = 0.6931$ that is, -1.6 dB. If the bandwidth is required not to be excessive, then this constraint is tightened to something like $\gamma_b > 0$ dB, the value of γ_S for $\kappa = 1$, and if the decoder must not be too elaborate, to something like $\gamma_b > 2$ or 3 dB.

The values of γ_b without coding depend on the permissible error-rate which with polar pulse-signalling is given by (4.52) and (4.20) as

$$P_e = \tfrac{1}{2}\operatorname{erfc}(\sqrt{\gamma_b}) \approx (4\pi\gamma_b)^{-1/2}\exp(-\gamma_b).$$

Thus for $P_e = 10^{-5}$ we find $\gamma_b = 9.6$ dB and for $P_e = 10^{-20}$ we find $\gamma_b = 16.3$ dB. Thus coding can save up to 7 dB if we can tolerate moderate error-rates and 13 dB if we want ridiculously low error rates (Exercise 5.18). On deep-space missions where every decibel is worth millions of dollars economic necessity demands efficient coding. But in more mundane situations the indications are not so strong.

(b) *Bandwidth-limited situation.* Let us now consider the obverse situation, with limited bandwidth and the power a secondary consideration, so that κ takes large values. This time the uncoded transmission employs multilevel real pulses

(Section 3.2.1) or complex pulses whose permitted amplitudes form a signal constellation (Section 3.4.4). To avoid discussing these two ways of doing things separately we may successively pair off the real-pulse amplitudes in a real-pulse signalling system and treat the pairs as complex amplitudes. We shall suppose that the real-pulse amplitudes are restricted to n equally spaced levels (Fig. 3.1(d) or Fig. 3.11(d)) spaced by μ so that the corresponding complex amplitudes form a square constellation of n^2 points (Fig. 3.11(f)). The bandwidth equivalent B (Section 5.1.2) is the number of complex amplitudes sent each second, and so the information rate J is related to the number of points $Q = n^2$ in the constellation by $Q = 2^{J/B} = 2^\kappa$. The mean energy per complex pulse is given by P/B, so that the mean energy available to each real pulse is $\frac{1}{2}P/B$. According to Section 3.2.1 this mean energy is $\mu^2(n^2-1)/12$. Thus it is found that

$$\tfrac{1}{2}P/B = \mu^2(n^2-1)/12 = \mu^2(Q-1)/12$$

and hence $\gamma_b = P/(\eta J)$ is given by

$$\gamma_b = \{\mu^2/(6\eta)\}\gamma_S \qquad (5.48)$$

with γ_S given by (5.47). Now the error probability is determined by the likelihood that the noise pushes the received level by more than $\mu/2$ towards an adjacent level (of which there are two except at the ends). Thus the error-probability is given by

$$P_W \approx 2 \cdot \tfrac{1}{2}\,\mathrm{erfc}\,(\tfrac{1}{2}\mu\eta^{-1/2}).$$

If a Gray code is in use, then such an error most likely causes only one bit to be wrong (Section 3.2.3). However each multilevel pulse conveys $\log_2 n = \tfrac{1}{2}\log_2 Q = \tfrac{1}{2}\kappa$ bits of information, and so the bit-error rate is

$$P_b = P_W/(\kappa/2).$$

A moderate error rate ($P_W = 1.05 \times 10^{-5}$) is obtained by setting $\mu^2/\eta = 39.0$. To obtain $P_b \approx 10^{-5}$ at $\kappa = 10$ we need $\mu^2/\eta = 33$. Thus by (5.48) we need to be something like 7.5 to 8.5 dB above the Shannon limit γ_S for a 'moderate' error-rate. We would expect to save a few dB by using error-correcting codes which can give comparable error rates at 2 or 3 dB above γ_S without difficulty. In most applications where the bandwidth is restricted there is little restriction on the power so that there is little need for coding.

On the other hand power is expensive on satellite links. This situation is exacerbated by the technical need to use angle modulation, with the signal-magnitude constant. The effect of this is roughly to halve the utilization of the bandwidth, a serious problem if the bandwidth is already at a premium (Exercise 5.19). Special codes have been developed for angle modulation, such as the multi-h phase codes (Section 6.2.3).

Finally it must be noted that although an increase in power is very effective

SHANNON'S THEOREM

in reducing error on an AWGN channel, it is not nearly so effective on channels with fading or with bursts of interference. Under these circumstances some form of coding must be used to achieve reliable communications.

Exercises

5.1. Show that the functions sinc(t) and sinc$(t - k)$ are orthogonal for any non-zero integral value of k. (Here the integral for the scalar product is from $t = -\infty$ to $+\infty$.)

5.2. If a basis-set is not orthogonal, then show that a set of vectors v_l satisfying $(v_l, u_k) = \delta_{lk}$ is given by $v_l = \Sigma_m c_{lm}^* u_m$, where the complex conjugates of the coefficients form a matrix $C = A^{-1}$ with A the matrix of coefficients $a_{kl} = (u_k, u_l)$, presumed non-singular.

5.3. Show that if two phasors satisfy Re$(\phi_A, \phi_B) = 0$, then the corresponding signals are orthogonal.

5.4. Show that the possible MSK signals (Section 3.4.5) over a given pulse-interval have phasors whose mutual scalar products have vanishing real parts. Hence use the result of Exercise 5.3 to show that the signals are orthogonal.

5.5. Two normalized vectors **u** and **v** are chosen at random in an M-dimensional Euclidean space. Show that the mean of $(\mathbf{u} \cdot \mathbf{v})^2$ is M^{-1}. (Hint: Consider **u** as fixed and **v** as chosen at random. Moreover let **u** be one of the vectors in an orthonormal basis set $\mathbf{u}_1, \mathbf{u}_2, \ldots \mathbf{u}_M$ with respect to which the components of **v** are $v_i = \mathbf{u}_i \cdot \mathbf{v}$. Then use $\Sigma v_i^2 = 1$.)

5.6. Data has to be sent at a rate of 9600 bits/s over an AWGN channel with a bandwidth equivalent of 2400 Hz. Show that a 16-point signal constellation will do the job. Estimate the minimum value of γ needed for $P_W < 10^{-5}$ if the following constellations are used: (a) 16-ASK, (b) 16-PSK, and (c) 16-QAM (Fig. 3.11(f)).

5.7. Show that if the p_l in (5.14) are orthonormal, then so are the u_k.

5.8. Show that if k is written as a binary number $(k_{m-1}, k_{m-2}, \ldots, k_0)$ with $k = 2^{m-1}k_{m-1} + 2^{m-2}k_{m-2} + \ldots + 2k_1 + k_0$ and similarly for l, then the elements of the matrix in Section 5.1.6 are given by

$$H_{kl} = (-1)^{k_0 l_0}(-1)^{k_1 l_1} \ldots$$

5.9. Prove by induction that any two rows of the $2^m \times 2^m$ matrix described in Section 5.1.6 differ in exactly $2^m/2$ places.

5.10. Compute the r_k of (5.15a) in the case $M = 8$, given that the q_l are $(0.5, -0.1, 0.2, -0.7, -0.1, 0.8, -0.1, 0.3)$. Which of the r_k has the largest magnitude?

5.11. Define the scalar product in (5.22) by (5.4) with T set to ∞. Assuming that the orders of averaging and of integration may be interchanged show that

$$\langle N_k^* N_l \rangle = \int_{-\infty}^{\infty} \int_{-\infty}^{\infty} a(t'-t'') u_k^*(t') u_l(t'') \, dt' \, dt'',$$

where $a(t'-t'') = \langle x^*(t') x(t'') \rangle$. Show that, for white noise with $A(f) = \frac{1}{2}\eta$, the autocorrelation function $a(t)$ is $\frac{1}{2}\eta\delta(t)$. Hence derive (5.23) using the orthonormality of the u_k.

5.12. Let S_p denote the event that s_p was transmitted, and R the event that a signal r was received. Use (4.2) to show that

$$P(S_p | R) = P(R) \cdot P(R | S_p) / P(S_p)$$

and hence that if $P(S_p)$ is independent of S_p then $P(S_p | R)$ is maximized with respect to S_p when $P(R | S_p)$ is maximized with respect to S_p.

5.13. Show that the minimum distance decoding strategy of Section 5.2.1 implies finding the signal point s_p which minimizes

$$\Sigma_{k=1}^{M} (s_{pk} - \rho_k)^2$$

where $s_{pk} = (u_k, s_p)$ and $\rho_k = (u_k, r)$.

5.14. Justify the algorithm described in Section 5.2.2 for finding the nearest lattice-point of D_4 to an arbitrary point R.

5.15. The compression factor of a lattice is defined as $1/(l_{min} \Delta^{1/d})$ where l_{min} is the minimum separation of points, d is the dimensionality, and Δ is the density of points. Show that this factor is always 1 for a simple cubic lattice in any number of dimensions and find it for (a) the two-dimensional hexagonal lattice, (b) D_4, (c) D_8, (d) E_8.

5.16. Find an upper bound on the error-probability for biorthogonal signalling by using the ordinary union bound, and show that it guarantees signalling with an arbitrarily low probability of error if $J' < \frac{1}{2} P/\eta$.

5.17. The coordinates X_i ($i = 1, \ldots, N$) of a point P in N dimensions are independent identically distributed random variables each with the probability density (4.16). Show that the probability for finding P lying between r and $r + dr$ from O is given by $\sigma(r)$ with $\sigma(r) = (\pi\eta)^{-N/2} A_N r^{N-1} \exp(-r^2/\eta)$ where A_N is the 'surface-area' of a hyper-sphere of unit radius in N dimensions. (a) Using

$$\int_0^\infty \sigma(r) \, dr = 1,$$

find an expression for A_N. (Assume N is even for simplicity.) (b) By considering $\ln \sigma$ show that σ is a maximum at $r = r_0$ with $r_0 = \sqrt{\{(N-1)\eta/2\}}$. (c) By expanding

$\ln \sigma$ as a Taylor series about r_0 show that for large N $\sigma(r)$ is in the vicinity of r_0 approximately a Gaussian with mean r_0 and variance $\eta/4$.

5.18. Find the mean time between erroneous bits (in years) on a channel sending 10^{10} bits/s if the bit-error rate is 10^{-20}.

5.19. In a narrow-band angle-modulated system an m-PSK signal constellation is used (Fig. 3.11(e)) with a *large* number m of points equally spaced around a circle. The magnitude of the spacing is μ. Show that

$$P/B = m^2 \mu^2/(4\pi^2)$$

with the notation of Section 5.3.7. Show also that (5.48) is replaced by the approximate result

$$\gamma_b \approx \{\mu^2/(2\pi^2 \eta)\} \gamma_S'$$

where

$$\gamma_S' = (2^{2\kappa} - 1)/(2\kappa).$$

(The comparison of this formula with (5.47) shows that γ_S' is in fact the Shannon lower bound on γ_b with a bandwidth equivalent $\tfrac{1}{2}B$, but with the same information rate J.)

Derive analogous formulae for an m-ASK carrier-based system with a constellation like that in Fig. 3.11(d).

5.20. Show that the decoding method of Section 5.2.4 applied to the PPM scheme of Section 4.3.4 is simply the strategy of finding which slot gives the largest amplitude. Describe a biorthogonal signalling system employing the Hadamard functions of Section 5.1.6 (with $M = 8$), and suggest a decoding scheme for it.

6. Error-correcting codes

This chapter is an introduction to the theory of error-correcting and error-detecting codes. The main aim in discussing such codes is to find efficient signalling systems on the AWGN channel. Reliable signalling over the AWGN channel has been discussed in Chapter 5, and the power and limitations of one particular form of signalling, orthogonal signalling, have been analysed. At the same time the opportunity will be taken to discuss other systems and other models for channels.

A major problem is finding efficient 'soft-decision decoding' algorithms which can process the real-valued outputs available from the demodulator when binary codes are sent as polar pulse-signals over an AWGN channel. These are contrasted with the 'hard-decision decoding' algorithms which can only process the binary values obtained after a 'hard decision' has been made on each pulse as to whether it is a 0 or a 1. Soft-decision algorithms for two simple binary block codes are discussed in Section 6.1, and the class of 'trellis' codes which can be decoded by the efficient Viterbi soft-decision decoding algorithm is discussed in Section 6.2. The theory of binary block codes is further developed in Section 6.3, in part to prepare the way for the next chapter. Soft-decision decoding of such codes and how well they perform on the AWGN channel are considered in Section 6.4.

6.1 Codes and channels

6.1.1 *The BSC and binary codes*

Let us suppose that a stream of bits (0 or 1) is sent over a binary channel which accepts as input a train of 0s and 1s and gives an output also of 0s and 1s which should be a faithful copy of the input. Errors can occur occasionally however with a 0 coming out as a 1 or vice-versa, and we wish to detect these errors or even better to correct them. Typically such a channel arises as follows. The input bit-stream is converted by a modulator into a polar pulsed signal which is then sent over an AWGN channel. At the receiving end a demodulator measures the amplitude of each pulse as received (Section 6.3.2). This amplitude is a real value. Then a *hard decision* is made by using the sign of the pulse-amplitude to decide whether a 0 or a 1 was sent.

Such a channel is a particular example of a Binary Symmetric Channel (BSC) which has the following properties:

(a) Both the input and the output streams are binary, that is, made up out of 0s and 1s.

(b) The probability of an error occurring with any given bit is independent of what has happened previously. Any channel with this property is called *memoryless*.

(c) The probability p of a 0 coming out as a 1 is the same as the probability of a 1 coming out as a 0. Such a channel is called *symmetric*.

In the case of the BSC produced as just described this error-probability p is given by (4.52) with E the energy available for each pulse and η the noise spectral density.

We now discuss some simple methods of error-detection and correction for a binary channel. In each case the encoder accepts the input stream in *blocks* of k bits, and it then puts in extra bits in each block as 'parity checks' to form a *codeword* of n bits to be sent over the binary channel. Such a code is called a '(n, k) binary code'. In this context the word 'bit' denotes an integer that only takes the values 0 and 1. A *bit-string* is a sequence of bits, and is denoted by a boldface letter. The bits in a bit-string **c** are denoted by c_i.

(a) *The single-parity-check code.* Perhaps the most familiar code of this sort is a *single-parity-check* (spc) code. A single extra bit c_{k+1} is appended to each block of k bits $(c_1, c_2, \ldots, c_k)$ so that the total number of 'set bits' (bits equal to 1) is even. This condition may be written as

$$c_1 + c_2 + \ldots + c_{k+1} = 0 \bmod 2 \tag{6.1}$$

where the phrase 'mod q' with q a positive integer signifies that any integer expression is to be brought back to the range 0 to $q - 1$ by adding or subtracting a suitable multiple of q. Thus an even (odd) number gives $0(1) \bmod 2$. Such a sum is called a 'parity check'. For example the block (1100111) (with $k = 7$) is extended to give (11001111). Every codeword is a string of $n = k + 1$ bits with an even number of 1s, and every string of $n = k + 1$ bits with an even number of 1s is a codeword. A single error on the BSC (or more generally an odd number of errors) is detected at the receiver because the number of 1s is now odd. The code is an $(n, n - 1)$ code, with $n = k + 1$.

(b) *The repetition code.* The single-parity check code is capable only of detecting errors. A very simple example of an error-correcting code is a *repetition code*. Suppose the message stream is broken up into blocks of length $k = 1$, and the encoder sends each bit three times, giving a $(3, 1)$ code. Then the possible codewords are (000) and (111). If a single error occurs it is corrected by using a majority vote. Thus (110) at the receiver would be interpreted by the decoder as (111), which corresponds to the single-bit message block (1). Similarly the decoder for a five-fold repetition code, with codewords (00000) and (11111) is capable of correcting any pattern of one or two errors, again by using a majority vote.

(c) *Array code.* In an array code we may picture the encoder arranging the bits of each message-block in a rectangular array, and then appending a parity-check bit to each row and column, so that the number of 1s in each row and column is even. Then the whole array is transmitted as a string. Table 6.1 shows a simple (8, 4) code of this type. Here the message-bits are c_1 to c_4 and there are four parity-check bits c_5 to c_8, determined by $c_1 + c_2 + c_5 = 0 \bmod 2$ and so on.

TABLE 6.1.

| c_1 | c_2 | c_5 |
c_3	c_4	c_6
c_7	c_8	

Thus for example let the message block be $(c_1 c_2 c_3 c_4) = (1011)$. Then the codeword $(c_1 c_2 \ldots c_8)$ is (10111001). It is obvious that any single error can be corrected. Thus if c_2 is wrongly received the parity checks on the first row and the second column fail and if c_6 is wrong the parity check on the second row fails, but nothing else.

(d) *A first-order Reed-Muller code*. As a final example of a binary code we consider the '(8, 4) first-order Reed-Muller code'. The encoder breaks the data into 4-bit message blocks which are then encoded into 8-bit codewords, in accordance with Table 6.2 (the grouping is done only to make the table more readable). The first eight codewords are simply the Hadamard functions of Section 5.1.6 modified by the prescription (3.4) that '+' is replaced by '0' and '−' by '1'. The remaining codewords are simply the negatives of the Hadamard functions similarly modified. This code can be shown to be able to

TABLE 6.2.

Input	Output
0000	0000 0000
0001	0101 0101
0010	0011 0011
0011	0110 0110
0100	0000 1111
0101	0101 1010
0110	0011 1100
0111	0110 1001
1000	1111 1111
1001	1010 1010
1010	1100 1100
1011	1001 1001
1100	1111 0000
1101	1010 0101
1110	1100 0011
1111	1001 0110

correct any single error, and to detect any pattern of two errors (Section 6.3.2). More generally we may use Hadamard functions to construct a whole family of (2^{k-1}, k) codes, the first-order Reed–Muller codes (Blahut 1983, Section 3.7). A decoder for the kth code can correct any pattern of $2^{k-3} - 1$ errors ($k \geqslant 3$), and detect any pattern of 2^{k-3} errors.

It is worth noting that error-correcting codes may be used purely for error-detection. In this case not only is the decoder simpler to construct, but also it is less easily fooled. For instance if the array code of Table 6.1 is used for correcting single errors, then a double error in bits c_6 and c_8 causes the decoder to change c_4, thus making things worse. On the other hand if the code is being used for error-detection, then parity-check failures warn the decoder that something is wrong.

It should also be noted that while binary codes have codewords made up from the alphabet $\{0, 1\}$, larger alphabets may also be used (Section 7.5).

Finally a mathematical point: equations like (6.1) involving integers evaluated mod 2 can be processed like ordinary equations (Section 7.1). There is an added bonus, that 'minus' may be replaced by 'plus', since $-a \doteq a$ mod 2 for any integer a. Thus for instance (6.1) implies the explicit result $c_{k+1} = c_1 + c_2 + \ldots + c_k$ mod 2.

6.1.2 Feedback and forward error-correction

In a two-way (duplex) communication system the receiver of a message may ask for a retransmission if part of the message has been garbled in transmission. This method of correcting errors is known as 'feedback error-correction', or Automatic Repeat reQuest (ARQ). In contrast the name 'forward error-correction' is given to a system of communication where errors in transmission are corrected at the receiving end without any request for retransmission. Thus a person listening to a distorted shortwave broadcast is using language as a forward error-correcting system. On the other hand on a bad telephone line the listener may ask the speaker to repeat himself, and so the system is using feedback error-correction.

On a two-way link feedback error-correction may well be better and simpler than forward error-correction. Here coding is used for error-detection and whenever an error is detected the receiver requests a retransmission. This makes things a lot easier at the receiving end, because error-detection is a much simpler operation than error-correction. Powerful codes may be used and the chances of an error getting through undetected can be made very low. Unfortunately repeats are not always convenient or even possible, as for instance on a broadcast channel, or on a recording, or in situations where the delay in propagation from transmitter to receiver is very long.

A third possibility is a mixture of feedback and forward-error correction where forward-error correction is normally used, and requests for retransmission are sent only when this fails. The receiver may even ask the transmitter over the feedback channel to vary the rate of transmission or the error-correcting scheme. This technique would be useful if the channel's characteristics changed slowly.

The emphasis in this chapter and in the next will be on forward error-correction, although feedback error-correction is discussed again in Section 6.4.4.

6.1.3 *Simple codes on the AWGN channel*

The codewords of a binary code may be sent over an AWGN channel as polar pulse-trains, set up according to the convention (3.4) say. In this way they become signals in a Euclidean signal space. We suppose that a pulse-string corresponding to an n-bit codeword

$$s(t) = \Sigma_{i=1}^{n} a_i p(t - iT_b) \qquad (6.2)$$

is sent out, where the $p(t - iT_b)$ are orthonormal pulse-functions (5.10), and where the coordinates a_i are given by

$$a_i = (1 - 2c_i) \cdot \sqrt{E} \qquad (6.3)$$

in terms of the codeword bits c_i (eqn 3.4). Here E is the energy of each pulse. The possible values of a_i are $\pm\sqrt{E}$. Because of the Gaussian white noise the measured values r_i of the a_i have added to them noise values x_i which are independent random variables each with the probability density function (4.16). The performance is better if the received pulse-values are treated as real numbers representing coordinates in this space, than if they are first converted back to bits by a hard-decision circuit. A decoder which works with real values rather than reconstituted bits is called a *soft-decision* decoder. Thus the $(n, n - 1)$ single-parity-check code has a modest error-correcting capability when used in this manner, rather than just a capability for error-detection (Section 6.4.1). We now describe such a decoding algorithm for this code.

A string of n received real values r_i has been measured by the demodulator and the nearest signal point in an n-dimensional Euclidean space is to be found. If there was no parity check these points would belong to the set of points with coordinates $\sqrt{E} \cdot (\pm 1, \pm 1, \ldots, \pm 1)$. The squared Euclidean distance may be regarded as a 'cost' to be minimized, which has the convenient property that it is the sum of 'costs', or terms, one for each received value. Thus the cost of rounding up (to $+\sqrt{E}$) the ith received value is then $(r_i - \sqrt{E})^2$, and of rounding it down is $(r_i + \sqrt{E})^2$. Evidently this is minimized by rounding to the nearer of $+\sqrt{E}$ and $-\sqrt{E}$. The cost of rounding the wrong way relative to the cost of rounding the right way is $4|r_i|\sqrt{E}$ and we may be forced to do this by the parity-check. Suppose that after rounding each received value to the nearer of $\pm\sqrt{E}$ we are unlucky and find an odd number of negative values. Then we have to choose at least one value to round the wrong way. Since the relative costs are all non-negative we choose just one of the r_i to round the wrong way and of course it is the one smallest in absolute value. (The similarity to the process of decoding lattice-codes (Section 5.2.2) should be noted.) Thus the decoding rule is simply to round all the n values and if this gives rise to a parity failure, to round the value smallest in absolute magnitude the wrong way. Thus this simple algorithm is indeed a minimum-Euclidean-distance algorithm.

Example. Suppose that in a system using the (5, 4) code the received values are (0.9, 0.2, −0.5, 0.3, 0.7). (We take $\sqrt{E} = 1$.) Rounding gives (1, 1, −1, 1, 1). Since this gives rise to a parity failure we round the value with the smallest magnitude the wrong way. This gives the signal point (1, −1, −1, 1, 1) which corresponds to the codeword (01100) and the message (0110).

Soft-decision decoding on an AWGN channel of two of the other codes from Section 6.1.1 is easily described. The $(n, 1)$ repetition code, which repeats each pulse $(n - 1)$ times, is not surprisingly equivalent to elongating the duration of each pulse by a factor n. Hence this code is hardly a code at all! The first-order Reed-Muller code when sent out over an AWGN channel becomes a method of biorthogonal signalling with Hadamard functions for the basis. Hence its decoding has already been described in Chapter 5, and may be carried out as described in Section 5.2.3 with the scalar products evaluated as in Section 5.1.6.

6.1.4 *Burst channels*

On a burst channel errors or erasures or both come in bursts, caused by bursts of interference affecting several consecutive symbols. Burst channels are not memoryless in the sense described in Section 6.1.1. Thus the Rayleigh fading channel (Section 4.4) often causes errors to occur in bursts.

It is evident that some form of coding must be used for such channels if reliable communications are to be assured, since increasing the power or slowing down the rate of transmission is not going to be very effective in increasing the reliability. Huge and impractical increases in power are needed to penetrate a fade, or a burst of interference caused by a pulsed source of interference close to the receiver. If coding is not used anything sent during the fade or burst is lost. Special codes for the forward-error correction of bursts of errors are available (Forney 1981). Naturally the duration of each codeword must be long in comparison with the burst-lengths, so that 'interpolation' across the bad patches is possible:

(a) There are special binary codes (such as the Fire codes) specially adapted for correcting bursts of erroneous bits within a codeword (Blahut 1983, Section 5.7).

(b) There are quasi-binary codes, such as the Reed–Solomon codes introduced in Section 7.5, where the codewords are made up from symbols in an alphabet of size 2^m ($m > 1$) and so are naturally themselves represented as strings of m bits. A burst of length m bits or less can at worst affect only two symbols.

128   ERROR-CORRECTING CODES

(c) The bits or symbols may be *interleaved*. After the encoder the symbols of a group of l consecutive codewords, each codeword being made up out of n symbols, are stored by rows in an array of l rows and n columns, one symbol in each location. Then the symbols are read out by columns and sent down the channel. This process is called interleaving to depth l. (It should be noted that, if as in the Reed-Solomon codes, each symbol is represented by several bits, the symbols themselves are not broken up in this process, only the codewords.) An error-burst that is not too long may knock out say one or two columns of this array, but only one or two symbols from each codeword are involved. (Interleaved Reed-Solomon codes are used in the Compact Disc digital audio system. See Hoeve, Timmermans, and Vries 1982.)

(d) The concept of interleaving can be developed to produce burst-error-correcting array codes (Farrell 1979).

(e) On radio channels where fading is very frequency-selective, so that one band of frequencies fades out while the neighbouring bands come through strongly, it may be worthwhile sending the symbols of a codeword simultaneously in parallel in different frequency-bands, rather than sequentially on the same frequency band.

6.2 Trellis codes and the Viterbi algorithm

6.2.1 *The duobinary code as a trellis code*

The trellis codes form a large family for which there is an efficient soft-decision decoding algorithm, the Viterbi algorithm (Bhargava, Haccoun, Matyas, and Nuspl 1981, Section 12.4). The reason for the name 'trellis' will become evident in a moment. The encoder performs a 'continuous flow' process, unlike the encoders for block codes, which encode each block of input of a specified length independently of the other blocks. To illustrate the idea we describe an encoder for the 'duobinary' code of Section 3.3.4. Although this code is usually employed for improving the frequency spectrum of a pulsed signal, it may also be used as a simple error-correcting code. Figure 6.1 illustrates the encoding. At the start of each step the encoder is in one of two states S_0 and S_1. A bit is then received from the source, and the encoder may then change its state by

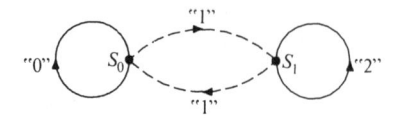

FIG. 6.1 Encoder for the duobinary code. Solid lines denote transitions caused by input 0, dashed lines by input 1. The symbols '0', '1', and '2' are the outputs.

TRELLIS CODES AND THE VITERBI ALGORITHM 129

following a solid line if the bit was '0' or by following a dotted line if the bit was '1'. At the same time the encoder puts one of a set of signals {0, 1, 2} as illustrated in the figure. It is important to note that the numbers in the figure refer to these outputs, not the inputs. At the beginning of the whole process the encoder is in the state S_0. What happens at the end is not very important but we shall suppose that the encoder returns to S_0 after N transitions altogether, of which $N - 1$ are determined by the message bits and the last one is chosen to return the encoder to S_0.

The history of the encoding process may be illustrated on a 'trellis', which is a $2 \times (N + 1)$ rectangular array of points joined by lines representing transitions. A trellis for the case $N = 6$ is shown in Fig. 6.2. The trellis point S_{ij} ($i = 0, 1$; $j = 0, 1, 2, \ldots, N$) represent the encoder as being in the state S_i at the jth stage.

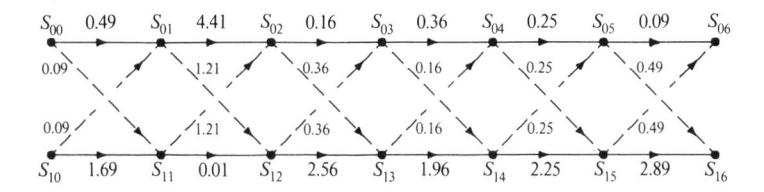

FIG. 6.2 Trellis for decoding duobinary signal. Solid lines are transitions associated with input 0, dashed lines with input 1. The decimal numbers give the partial cost of each transition, given that (0.7, 2.1, 0.4, 0.6, 0.5, 0.3) was received.

As before a solid line represents a transition caused by an input 0 and a dotted line a transition caused by 1. Suppose that the encoder is given the message 10111 (reading from left to right). This then causes the encoder to follow a path through the trellis from S_{00} to S_{11} to S_{12} to S_{03} to S_{14} to S_{05} and finally to S_{06}. The output sequence is 121110.

This coded signal is then sent over a noisy channel. Let us suppose that it is received as the sequence 0.7, 2.1, 0.4, 0.6, 0.5, 0.3. The decoder has to decide which of the $2^{N-1} = 32$ possible sequences was put into the encoder, or equivalently which of $2^{N-1} = 32$ paths was followed through the trellis. The criterion for the choice will be the path of minimum 'cost', the cost being given by the sums of the squares of the differences between the received values and the values of the outputs corresponding to that path. Thus for the path $S_{00}S_{11}S_{12}S_{13}S_{14}S_{15}S_{06}$ which corresponds to the coded signal 122221 and to the message 10000 the squares of the difference are respectively 0.09, 0.01, 2.56, 1.96, 2.25, 0.49 with a total cost of 7.36. We then try all paths and choose the one with the minimum cost. For each possible transition in the trellis there is a corresponding output 0, 1 or 2 as given by Fig. 6.1, and the numbers in Fig. 6.2 give the costs of squares of the differences from the corresponding received values. These costs are often referred to as 'metrics'.

6.2.2 *The Viterbi algorithm*

The brute-force approach of trying each of the 2^{N-1} possible paths from S_{00} to S_{0N} may be feasible when $N = 6$, but not when $N = 1000$. The Viterbi algorithm builds up minimum-cost paths from stage to stage with a computing effort proportional to N. The basic idea is very simple. The minimum-cost path to some point $S_{j,n+1}$ in the trellis must have passed through one of the preceding points S_{in}. The part from S_{00} to S_{in} must be a minimum-cost path from S_{00} to S_{in}, for otherwise the overall cost could be lowered by changing this part appropriately. Thus minimum-cost paths are built up by extending shorter minimum-cost paths. The method will be illustrated with the costs shown in Fig. 6.2.

The minimum-cost path to S_{02} is evidently $S_{00}S_{11}S_{02}$ with a cost of 1.3 and to S_{12} it is $S_{00}S_{11}S_{12}$ with a cost of 0.1. Each of these is chosen from two alternatives. The minimum-cost path to S_{03} is either $(S_{00}S_{11}S_{02})S_{03}$ or $(S_{00}S_{11}S_{12})S_{03}$, the brackets showing previously found minimum-cost paths. The costs are respectively $1.3 + 0.16 = 1.46$ and $0.1 + 0.36 = 0.46$. Obviously the minimum-cost path to S_{03} is $S_{00}S_{11}S_{12}S_{03}$ with a cost of 0.46. Similarly we find that the minimum-cost path to S_{13} is $(S_{00}S_{11}S_{02})S_{13}$ with a cost of $1.3 + 0.36 = 1.66$ rather than $(S_{00}S_{11}S_{12})S_{13}$ with a cost of 2.66. In this way the minimum-cost path from S_{00} to S_{06} is found step-by-step (Exercise 6.3).

There is one practical problem in the algorithm as described when N is large. The decoder has to store the minimum-cost paths back to the start, and so a lot of storage is needed. Moreover there is no output until the very end. Fortunately in all reasonable cases the minimum-cost paths are coalesced several stages back. Thus for the third stage in our example the two minimum-cost paths $S_{00}S_{11}S_{12}S_{03}$ and $S_{00}S_{11}S_{02}S_{13}$ are coalesced from S_{00} to S_{11}. Thus we may put out the decoded bit 1 corresponding to the transition from S_{00} to S_{11} and use S_{11} instead of S_{00} as the starting point. This happens in the general case with a good code because received values at some later time will not influence decisions about the decoded signal at a much earlier time.

It is evident that the family of trellis codes is very large. The number of rows in the trellis may be any integer larger than 1, and the input to the encoder need not be bits but symbols from a larger alphabet, so that there are many lines coming out of each stage. There are many ways of choosing the output signals. The important thing is that the cost or metric should be additive. Thus for instance if we are trying to maximize a product of conditional probabilities when using maximum-likelihood decoding on a memoryless channel, then the obvious cost-function is the negative of the logarithm of these probabilities. In the next two sections we give examples of important trellis codes.

6.2.3 *MSK modulation and the multi-h phase codes*

For satellite communications at microwave frequencies angle modulation is far more convenient than amplitude modulation because transmitter output tubes

TRELLIS CODES AND THE VITERBI ALGORITHM   131

operate best at constant power. One simple form of binary digital modulation is as follows: The modulating phasor has the form $\phi(t) = A \exp\{j\theta(t)\}$, where A is a constant. During the time T_b assigned for each bit the phase $\theta(t)$ changes linearly by $+\frac{1}{2}\pi$ if the bit is 0 and by $-\frac{1}{2}\pi$ if the bit is 1. Thus $\theta(t)$ and hence $\phi(t)$ are continuous functions of t so that the Fourier transform $\Phi(f)$ of $\phi(t)$ falls off fairly rapidly for large f, in fact as f^{-2} since the derivative is discontinuous. At the beginning of each interval there are four possible states S_m of the encoder, corresponding to the four different values for $\theta(t)$ of $\frac{1}{2}m\pi$ ($m = 0, 1, 2, 3$). If a 0 comes in, then $\theta(t)$ changes by $+\frac{1}{2}\pi$ and the final state is S_n where $n = m + 1 \bmod 4$. The output is simply $\phi(t)$ (or strictly a carrier modulated by $\phi(t)$). With a received signal phasor $r(t)$ the decoder finds that possible transmitted $\phi(t)$ which minimizes $\int |\phi(t) - r(t)|^2 \, dt$ integrated over the duration of the signal. It is evident that this cost-function can be broken up as required into a sum of partial costs, one for each transition.

This is an example of a Continuous Phase Shift Keying (CPSK) system. It is in fact the MSK system mentioned in Section 3.4.5.

One trouble with the MSK system is that it is possible for two trellis-paths to coalesce back to the same state after only two transitions. This reduces the error-correcting capability of the system. Thus the inputs 00 and 11 both take the encoder from an initial state S_0 to S_2. *Multi-h phase-codes* were developed as a means of improving this situation. Here is a typical example of such a code:

The phase-angle $\theta(t)$ of the phasor can take the values $2\pi m/16$ ($m = 0, 1, \ldots, 15$), so that the encoder is in one of 16 states S_m. When the first bit comes in the value of m is changed to $m \pm 6 \bmod 16$, the sign depending on the bit. When the second bit comes in the value of m is changed by $\pm 5 \pmod{16}$, and when the third bit comes in by $\pm 4 \pmod{16}$. The fourth bit changes m by ± 6 again, and so on. It is readily seen that it is impossible to attain the same state by two different paths in less than four transitions. Figure 6.3 shows the eight possible values of m achievable after three transitions if the encoder starts

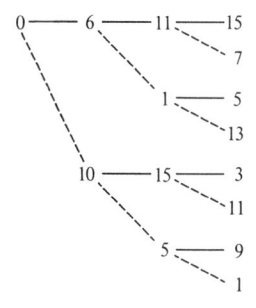

FIG. 6.3 Phase-states achieved in three transitions of a multi-h phase code. Solid lines are transitions associated with input of 0 to encoder, dashed lines with input of 1.

with $m = 0$. Thus the input 011 takes us to the state S_{13} and 111 to S_1. At the fourth stage the paths can coalesce. Thus 0111 takes us to S_7 (via S_{13}), as does 1110 (via S_1).

The reason for the name is as follows: The h-value in a CPSK system is the magnitude of the phase-change during a bit-interval, as a multiple of π. Thus in the MSK system we have $h = \frac{1}{2}$. The multi-h phase-codes vary the h-value with the bit-interval in a periodic way. In the above example the h-values are $\frac{6}{8}, \frac{5}{8}$, and $\frac{4}{8}$, and then they repeat.

Calculating the error-probabilities and finding the best codes is quite a complicated business (Bhargava et al. 1981, Chapter 6).

6.2.4 Convolutional codes

A simple convolutional encoder is shown in Fig. 6.4. It consists of a 3-stage shift-register and two adders (mod 2). At the start of any stage the next bit from the source is fed into the right of the shift-register, pushing the left-hand bit out. Then the sum mod 2 of the first and third bits is sent out, followed by the sum mod 2 of all three bits. Hence the output consists of bit-pairs. It may seem that the encoder can be in any one of eight possible states, but since at the start of every transition the left-hand bit is discarded without doing anything there are in effect only four possible states determined by the right-hand pair of bits.

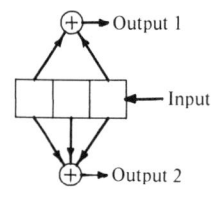

FIG. 6.4 Convolutional encoder.

The reason for the name 'convolutional' is as follows: Suppose that the input bit stream is $m_0 m_1 m_2 \ldots$ and the output $a_0 b_0 a_1 b_1 a_2 b_2 \ldots$ Then these are related by $a_i = m_i + m_{i-2}$, $b_i = m_i + m_{i-1} + m_{i-2}$ $(i = 0, 1, 2, \ldots)$ where m_{-1} and m_{-2} are taken as zero. These equations may be written in the form:

$$a_i = \Sigma_j g_j m_{i-j} \bmod 2,$$

$$b_i = \Sigma_j h_j m_{i-j} \bmod 2$$

with the g_j and h_j zero except for g_0, g_2, h_0, h_1, h_2 which are 1. These equations have the form of additive convolutions like those in Section 2.3.1 (eqn 2.44).

More powerful convolutional codes have longer shift-registers. If the shift-register is K bits long, then the number of states M is 2^{K-1} and the decoding complexity grows exponentially with K. The quantity K is called the 'constraint length'. The largest practical values of K are in the range 10 to 12.

Bounds on the error probability of a given code are quite readily found by techniques described by McEliece (1977). The best codes are found by analysing the error probabilities in this manner. Typically convolutional codes with a constraint of length K are better than block codes with K information bits per block, and as with block codes the error probability falls exponentially with K. Unfortunately we have just seen that as with minimum-distance decoding algorithms for block codes the decoding complexity also grows exponentially with K.

6.3 Binary linear block codes

6.3.1 *Introductory*

The theory of binary block codes introduced in Section 6.1.1 is now developed further. *Hamming weight* and *Hamming distance* are discussed first. The Hamming weight $w(\mathbf{u})$ of a bit-string $\mathbf{u}$ is simply the number of 1s in it, and the Hamming distance $d(\mathbf{u}, \mathbf{v})$ between two bit-strings $\mathbf{u}$ and $\mathbf{v}$ of the same length n is the number of places in which they differ. These quantities are ordinary integers. Thus the Hamming weight of the bit-string (110110) is 4 and the Hamming distance between the bit-strings (110110) and (100111) is 2, since they differ in the second and sixth places. The minimum distance of a code is the minimum Hamming distance between codewords.

The Hamming distance obeys the so-called triangle inequality:

$$d(\mathbf{u}, \mathbf{w}) - d(\mathbf{v}, \mathbf{w}) \leq d(\mathbf{u}, \mathbf{v}) \leq d(\mathbf{u}, \mathbf{w}) + d(\mathbf{v}, \mathbf{w}). \tag{6.4}$$

To prove this we first note that

$$d(\mathbf{u}, \mathbf{v}) = \Sigma_{i=1}^{n} d(u_i, v_i) \tag{6.5}$$

with

$$d(u_i, v_i) = |u_i - v_i|. \tag{6.6}$$

Here n is the number of bits u_i and v_i in the strings $\mathbf{u}$ and $\mathbf{v}$ respectively. These numerical differences are readily shown to obey the triangle inequalities and hence we obtain (6.4) using term-by-term summation.

The sum or difference $\mathbf{c} = \mathbf{a} \pm \mathbf{b}$ of two bit-strings $\mathbf{a}, \mathbf{b}$ is defined by

$$c_i = a_i \pm b_i \bmod 2.$$

However, since $a_i - b_i = a_i + b_i \bmod 2$, because the two sides differ by a multiple of 2, the difference $\mathbf{a} - \mathbf{b}$ is the same as the sum $\mathbf{a} + \mathbf{b}$. For example the sum (or difference) of the bit-strings (110110) and (100111) is (010001), with 0 in the answer where corresponding bits are the same and 1 where they are different. Hence it follows that

$$d(\mathbf{u}, \mathbf{v}) = w(\mathbf{u} - \mathbf{v}) = w(\mathbf{u} + \mathbf{v}). \tag{6.7}$$

If $\mathbf{v}$ is set to $\mathbf{0}$, where $\mathbf{0} = (0, 0, \ldots, 0)$ is the string of zeros, it is found that

$$d(\mathbf{u}, \mathbf{0}) = w(\mathbf{u}), \tag{6.8}$$

and thus setting $\mathbf{w} = \mathbf{0}$ in (6.4) gives

$$w(\mathbf{u} - \mathbf{v}) \leqslant w(\mathbf{u}) + w(\mathbf{v}). \tag{6.9}$$

(The symbol '+' here denotes ordinary addition.)

A *binary linear code* is one where the codewords are bit-strings of a specified length, and where the sum of any two codewords is also a codeword. All the codes described in Section 6.1.1 are linear codes, as may be readily verified. In consequence we find that $\mathbf{0} = (0, 0, \ldots, 0)$ is always a codeword, being the sum of a codeword with itself, and by (6.7) that the minimum distance $d_{\min}$ of the code (i.e. the minimum distance between codewords) is the minimum of the weights of the non-zero codewords. Thus the minimum distance of the parity check code is 2 and of the (8, 4) Reed–Muller code is 4, as may be verified from a list of its codewords.

It is often very convenient to arrange that there is a linear relation between the message $\mathbf{m}$ and the codeword $\mathbf{c}$ of the form

$$c_j = \Sigma_{i=1}^{k} g_{ji} m_i \bmod 2, \qquad j = 1, 2, \ldots, n. \tag{6.10}$$

The $n \times k$ matrix g_{ji}, whose elements are either 0 or 1, is called the *generator matrix*. This property does not affect the error-correcting capabilities of the code, at least not if all the messages are equally likely, but it can greatly simplify the encoding and decoding processes. In all the codes described in Section 6.1.1 the message blocks are assigned to the codewords in this manner.

If the message bits appear in specified locations of the codeword then the code is called *systematic*. Otherwise it is called non-systematic. (Strictly this is a property of the assignment of the message-blocks to the codewords, rather than of the code itself.) The codes in Section 6.1.1 other than the Reed–Muller code are systematic.

6.3.2 *Binary hard-decision coding*

The single-parity-check code and the first-order Reed–Muller code have straightforward soft-decision decoding algorithms, but unfortunately this is not true for most binary linear codes. Many of these codes have simple decoding algorithms which can only handle binary digits. In consequence the decoder has to be preceded by a *hard-decision* circuit, which uses the sign of each received pulse-amplitude to decide whether a 0 or 1 was sent (Section 4.3.3). The error probability on the AWGN channel has already been shown to be (4.52)

$$p = \tfrac{1}{2}\mathrm{erfc}\{\sqrt{(E/\eta)}\} \quad \text{so that} \quad p = \tfrac{1}{2}\mathrm{erfc}\{\sqrt{(\gamma_b R)}\}. \tag{6.11}$$

Here
$$R = k/n \tag{6.12}$$
is the *rate* of an (n, k) code, and γ_b is given by
$$\gamma_b = E_b/\eta, \tag{6.13}$$
the ratio of the energy E_b available for each bit of information to the noise energy η. Here E_b is given by $nE = kE_b$, so that
$$E_b = nE/k. \tag{6.14}$$

The channel is now in effect a binary symmetric channel. If all messages are regarded as equally likely, then as in Section 5.2.1 maximum likelihood decoding (MLD) is obtained by answering the question 'Which codeword makes the bit-string r that was received the most likely?' If we consider a particular codeword c that differs from r in d places (and agrees with r in the $(n-d)$ remaining places), then the probability that the channel changes c into r is

$$p^d(1-p)^{n-d} = (1-p)^n \{p/(1-p)\}^d.$$

(There is no combinatorial factor here since this is the probability of a particular pattern of errors, rather than the probability of any pattern of d errors.) It is evident that this probability falls with increasing d, provided that $p < \frac{1}{2}$, which follows from (6.11). Hence the codeword to choose is the one which makes d the smallest. Thus minimum-Euclidean-distance decoding is replaced by minimum-Hamming-distance decoding.

Let us see how this works for the (8, 4) Reed–Muller code. We find which of the 16 codewords has the minimum Hamming distance from the received bit-string r. A single error means that r is at a Hamming distance of 1 from the correct codeword c, but it must be at a distance of at least 3 from any other codeword by the inequality (6.4) and by the fact that the Hamming distance between codewords is at least 4. Thus a single error is always corrected. A double error leaves the received bit-string at a distance 2 from the correct codeword, but it may also be at a distance 2 from some other codeword as well. So all double errors are detected but not necessarily corrected.

More generally if the number of errors in the received word is less than or equal to the integer

$$t = \text{int} \{\tfrac{1}{2}(d_{\min} - 1)\} \tag{6.15}$$

then the above algorithm always returns the correct codeword. This is easily seen as follows: Let c and r denote the transmitted codeword and the received word respectively, and let c′ denote another codeword also at a Hamming distance from r less than or equal to t. Then it follows that

$$d(c, c') \leq d(c, r) + d(c', r) \leq 2t \leq d_{\min} - 1 < d_{\min}.$$

The first inequality is the triangle inequality (6.4). So these codewords are closer than $d_{\min}$, which contradicts the definition of $d_{\min}$ as the minimum distance.

We call t the number of errors that the code can correct, and the code a t-error correcting code. (Minimum-Hamming-distance decoding may work with some patterns of more than t errors, provided that these leave the received word **r** closer to the correct codeword **c** than any other codeword.)

An upper bound on the error probability for minimum-Hamming-distance decoding is readily obtained. (Error detection without correction is here considered an error.) The probability P_E of an error occurring is certainly less than or equal to the probability of having more than t erroneous bits, and so the argument leading up to (4.1) in Section 4.1.1 shows that

$$P_E \leq \Sigma_{s=t+1}^{n} \binom{n}{s} p^s (1-p)^{n-s}. \tag{6.16}$$

As a function of γ_b the asymptotic behaviour of P_E for large γ_b is

$$P_E \sim \exp(-(t+1)R\gamma_b). \tag{6.17}$$

This result is obtained by using the approximation (4.21) in (6.11), after which only the lowest power of p in (6.16) is kept. The combinatorial coefficient $\binom{n}{t+1}$ is a constant, independent of γ_b.

We list some of the many types of decoding algorithm for the BSC:

(a) *Minimum-Hamming-distance algorithm.* Each codeword is examined to see which is the nearest the received word (in Hamming distance). This is readily implemented for the Reed–Muller code, and the computations can be speeded up by adapting the method of Section 5.1.6. Instead of real numbers the inputs are the integers 1 for bit 0 and −1 for bit 1. The algorithm is good for low-rate codes where there are relatively few codewords in comparison with the number of possible bit-strings of the same length.

(b) *Syndrome decoding.* This will be described in Section 6.3.3. It is best for high-rate codes.

(c) *Algebraic decoding.* This can be used for certain codes. It is described in detail in Chapter 7. It is not unlike the techniques used in numerical analysis for finding errors in tables of values or for interpolating missing values. It is both a limited technique in that there is no soft-decision equivalent and it is also very powerful, because the decoding complexity goes as the square of the codeword length (at a fixed rate) rather than exponentially, unlike all the other methods described. It thus opens up the practical possibility of using highly reliable codes with large values of n, of the order of thousands if required. It will be seen in Chapter 7 how this technique may be combined with soft-decision decoding to give very powerful codes, which overcome the complexity problem found with orthogonal signalling, with Shannon's random codes, and with Viterbi decoding.

(d) Other methods may be found in Clark and Cain (1981). A technique called majority-logic decoding, applicable only to certain codes, is exemplified in Section 6.3.5.

It should finally be noted that some of these algorithms are incomplete in the sense that they cannot correct all the error-patterns that can in principle be corrected by a minimum-distance algorithm.

6.3.3 *The parity-check matrix and the syndrome*

The number of parity-check bits in an (n, k) code is equal to $(n-k)$ and so may be specified by $(n-k)$ equations. These equations can be written as

$$\sum_{j=1}^{n} h_{ij} c_j = 0 \bmod 2, \quad i = 1, 2, \ldots, (n-k). \tag{6.18}$$

Here the codeword bits c_j are labelled from 1 to n and the equations from 1 to $n-k$. The h_{ij} are equal to 0 or 1 depending on whether c_j appears in the ith equation. Thus the parity-check equations akin to (6.1) for the (8, 4) array code with appropriate ordering give the matrix shown in Table 6.3.

TABLE 6.3

$i \setminus j$	1	2	3	4	5	6	7	8
1	1	1	0	0	1	0	0	0
2	0	0	1	1	0	1	0	0
3	1	0	1	0	0	0	1	0
4	0	1	0	1	0	0	0	1

Any bit-string c of length n satisfying (6.18) is a codeword, and every codeword satisfies (6.18). The code must be linear since if **a** and **b** are codewords then so is **c** = **a** + **b**. For we see that

$$\sum_j h_{ij} c_j = (\sum_j h_{ij} a_j) + (\sum_j h_{ij} b_j) = 0 + 0 = 0 \bmod 2$$

using (6.18) on **a** and **b**. Hence any code defined by means of its parity checks must be a linear code. The parity-check matrix may be used to define a code, since it specifies how the parity-checks are set up. The parity-check matrix of the single-parity-check code is just a row of 1s. An equivalent parity-check matrix may be formed by adding (mod 2) any row to any other, so the parity-check matrix of a given code is by no means unique.

Suppose in transmission certain bits in **c** are changed. These changes can be simulated by adding a bit-string **e** of errors to **c** where the non-zero bits of **e** mark the changes. The parity checks are

$$s_i = \sum_j h_{ij}(c_j + e_j) = \sum_j h_{ij} c_{ij} + \sum_j h_{ij} e_j = 0 + \sum_j h_{ij} e_j = \sum_j h_{ij} e_j \tag{6.19}$$

(all mod 2), and the collection of parity checks is called the 'syndrome'. Thus in the array code of Section 6.1.1, if c_2 is received incorrectly then **e** = (01000000) and the syndrome is $s_1 = 1$, $s_2 = 0$, $s_3 = 0$, and $s_4 = 1$, or **s** = (1001) when written as a string. The syndrome exactly matches the second column of h_{ij} and thus it locates the error. This is because with a single error in the pth location we find $s_i = h_{ip}$ from (6.19). Some double errors are detected because they give syndromes which cannot be matched in this way (say errors in c_3 and c_5) but others give a syndrome which leads to a false correction (say errors in c_1 and c_5, which cause c_7 to be changed as well).

In the case of single-error correcting codes the syndrome immediately determines the error, if of course there is only one error. In more elaborate multiple-error correcting codes the syndrome may be used as a pointer to a look-up table giving the corresponding error-pattern (of least weight) which can be corrected (syndrome decoding). This is because the syndrome for every error pattern of weight $<\frac{1}{2}d$ (where d is the minimum Hamming distance of the code) does not correspond to any other error pattern of weight $<\frac{1}{2}d$. For suppose $H\mathbf{e} = \mathbf{s}$ and $H\mathbf{f} = \mathbf{s}$ with $w(\mathbf{e}) < \frac{1}{2}d$ and $w(\mathbf{f}) < \frac{1}{2}d$. Then $H(\mathbf{e}-\mathbf{f}) = 0$. So $\mathbf{e} - \mathbf{f}$ is a codeword and hence $w(\mathbf{e} - \mathbf{f}) \geqslant d$. Hence by (6.7) we obtain a contradiction. Thus every error-pattern of weight less than $\frac{1}{2}d$ is identified by the syndrome. (It can be shown that every syndrome corresponds to some error-pattern, but not necessarily of weight $<\frac{1}{2}d$, and that for syndromes corresponding to error-patterns all with weights $\geqslant \frac{1}{2}d$ the minimum-weight error-pattern may not be unique.)

The number of bits in the syndrome is equal to the number of parity-checks which in turn is equal to $n - k$ where k is the number of message-bits and n is the number of bits in a codeword. Thus the number of possible syndromes is 2^{n-k} and so a look-up table is practicable only if $n - k < 15$, which usually means that the code is a high-rate code. As will be described in the next chapter the syndrome may also be used as the input to an algebraic decoding algorithm where the restriction on $n - k$ is much weaker.

6.3.4 *Hamming codes*

Hamming codes have codewords of length $n = 2^m - 1$ for any positive integer value of m and k equal to $n - m$, so that there are m parity-check equations. The parity check matrix may be written simply as the integers 1 to $2^m - 1$ expressed in binary and set down as the columns. Thus setting $m = 3$ gives a (7, 4) code with parity check matrix

$$\begin{bmatrix} 0 & 0 & 0 & 1 & 1 & 1 & 1 \\ 0 & 1 & 1 & 0 & 0 & 1 & 1 \\ 1 & 0 & 1 & 0 & 1 & 0 & 1 \end{bmatrix}.$$

These codes are single-error correcting because the syndrome of an error in the ith location is simply the ith column of the parity check matrix H, and this is just the integer i in binary. Double errors may lead to false corrections, in fact they always do, and the minimum distance is 3.

6.3.5 Two particular codes

Two multiple error-correcting codes are now discussed. The first one is a (15, 7) double-error-correcting code with $d_{min} = 5$ and the second the important (23, 12) triple-error-correcting Golay code with $d_{min} = 7$ (MacWilliams and Sloane 1977). The 8 × 15 parity-check matrix for the (15, 7) code may be written as in Table 6.4 with the unmarked locations being zero. (The columns are numbered from 1 to 15.) The first eight bits may be taken as parity checks, since the last equation can be used to express c_8 in terms of c_9 to c_{15}, the penultimate can be used to express c_7 in terms of c_9 to c_{15} as well after eliminating c_8, and so on. There are 128 codewords and it may be verified that the end-around cyclic shift of any codeword is also a codeword.

TABLE 6.4

$$\begin{bmatrix} 1 & 1 & 0 & 1 & 0 & 0 & 0 & 1 & & & & & & & \\ & 1 & 1 & 0 & 1 & 0 & 0 & 0 & 1 & & & & & & \\ & & 1 & 1 & 0 & 1 & 0 & 0 & 0 & 1 & & & & & \\ & & & 1 & 1 & 0 & 1 & 0 & 0 & 0 & 1 & & & & \\ & & & & 1 & 1 & 0 & 1 & 0 & 0 & 0 & 1 & & & \\ & & & & & 1 & 1 & 0 & 1 & 0 & 0 & 0 & 1 & & \\ & & & & & & 1 & 1 & 0 & 1 & 0 & 0 & 0 & 1 & \\ & & & & & & & 1 & 1 & 0 & 1 & 0 & 0 & 0 & 1 \end{bmatrix}$$

Error-correction may be carried out by using a syndrome lookup table with $2^8 = 256$ entries, but there is an interesting alternative method of locating single or double errors. Out of the 8 components in $s = H(c + e) = He$ the following are selected:

$$s_1 = e_1 + e_2 + e_4 + e_8,$$

$$s_5 = e_5 + e_6 + e_8 + e_{12},$$

$$s_7 = e_7 + e_8 + e_{10} + e_{14},$$

$$s_8 = e_8 + e_9 + e_{11} + e_{15}.$$

All four equations contain e_8, while the other components of e appear at most once. If there are one or two errors, then at most two bits in e are non-zero. If e_8 is 1, then at least three of these syndrome bits are 1, but if e_8 is 0 then at most two are 1. Thus it is evident whether e_8 is in error. But since the end-around shift of a codeword is also a codeword each bit in turn may be brought into the eighth location for testing. This technique is known as 'majority logic decoding' (Blahut 1983) and is unfortunately only available for a very restricted range of codes.

That double errors can always be corrected shows that $d_{min} \geqslant 5$, but it is not obvious that $d_{min} = 5$. However it would not be hard to program a computer to verify this. This code is one of the family of cyclic codes to be considered in Chapter 7.

This code may be *extended* to a (16, 7) code with $d_{min} = 6$ by putting in an overall parity check. The extended code is still double-error correcting, but it will now also detect all patterns of three errors.

The (23, 12) Golay code has a 11 × 23 parity check matrix not unlike the one of the code just described. The rows are given by the string

$$10\ 100\ 100\ 111\ 110\ 000\ 000\ 000$$

and its 10 cyclic shifts to the right. This code may also be extended to a (24, 12) code with $d_{min} = 8$ by means of an overall parity check. A cunning 'error-trapping' decoding algorithm attributed to Berlekamp for the extended code is described by MacWilliams and Sloane (1977). Unfortunately it would take too long to describe here. However the code can also be decoded by means of a syndrome lookup table with $2^{12} = 4096$ entries, and such a decoder is not hard to implement with modern semiconductor memories.

6.4 Binary block codes on the AWGN channel

6.4.1 *Binary linear codes on the AWGN channel*

Binary codes readily provide signal-space codes, as was shown in Section 6.1.3. An orthonormal set of functions $u_i(t)$ ($i = 1, \ldots, n$) may be used to set up a signal representation of any codeword c from a given (n, k) code as

$$s(t) = \sqrt{E} \cdot \Sigma_{i=1}^{n}(1 - 2c_i)u_i(t)$$

The total energy is nE, or E per bit of codeword {eqns (6.2), (6.3)}. Here the c_i are treated on the right-hand side as ordinary numbers. Thus the square d_E^2 of the Euclidean distance between the signals s, s' corresponding respectively to the codewords c, c' with a Hamming distance d is $4Ed$. This is because

$$d_E^2 = (s - s', s - s') = E\Sigma_{i=1}^{n}\{(1 - 2c_i) - (1 - 2c_i')\}^2$$
$$= 4E\Sigma_i(c_i - c_i')^2 = 4E\Sigma_i|c_i - c_i'| = 4Ed.$$

The second equality follows from (5.13), the fourth from $c_i, c_i' = 0$ or 1, and the last from (6.5) and (6.6). The signals form a spherical code in signal space with radius $\sqrt{(nE)}$, since they all have the same energy. A few points are worthy of note:

(a) In a linear code the number of codewords at a given Hamming distance d from a given codeword c is the same as the number of codewords at distance d from the all-zeros codeword, so it is just A_d, the number of codewords of weight d, and is independent of the choice of c. Hence the error probability is the same no matter which codeword is sent, and it is usually simplest to consider the error-probability in the case that the all-zeros codeword is sent.

BINARY BLOCK CODES ON THE AWGN CHANNEL

(b) The error probability can be bounded by the union bound. Such a result is good when the signal-to-noise ratio is high. It is found that

$$P_W < \Sigma_{d(>0)} A_d \cdot \tfrac{1}{2} \operatorname{erfc}\{\sqrt{(\gamma_b R d)}\} \qquad (6.20)$$

(Exercise 6.13), and hence by (4.22) it follows that

$$P_W < \tfrac{1}{2} \operatorname{erfc}\{\sqrt{(\gamma_b R d_{\min})}\} \Sigma_d A_d \exp\{-\gamma_b R(d - d_{\min})\}. \qquad (6.21)$$

where R and γ_b are given by (6.12) and (6.13). The rough approximation (4.21) shows that P_W behaves as

$$P_W \sim \exp(-\gamma_b R d_{\min}) \qquad (6.22)$$

for large γ_b. Tighter results, useful for small γ_b, may be obtained by the method of the tangential-union bound (Section 5.2.4).

(c) It is helpful to picture the u_i as quasi-rectangular non-overlapping pulse-functions, as indeed they may well be.

The rather rough concept of *coding gain* can now be introduced. It is the ratio of the energy per bit needed to attain a specified low error-probability (say $P_W = 10^{-5}$) without using the code to the energy per bit needed to attain the same error-probability using the code. Like any ratio of energies it is usually expressed in dB. A rough value may be obtained as follows: In the uncoded case it is found that

$$P_W \sim \exp(-\gamma_b')$$

where γ_b' is the energy per bit for this case divided by η. This result is obtained by setting $d_{\min} = 1$ and $R = 1$ in (6.22). Again by (6.22) this is equal to P_W for the coded case if

$$\gamma_b' \approx \gamma_b R d_{\min}.$$

Thus the gain is roughly $R d_{\min}$.

As an example let us consider the $(n, n-1)$ single-parity-check code on the AWGN channel (Section 6.1.3). Here $R = (n-1)/n$ and $d_{\min} = 2$, so that the coding gain is just under 2. This is quite a useful gain, especially in view of the simplicity of the code and of its decoding algorithm.

6.4.2 Performance of Reed–Muller codes

The first-order Reed–Muller code described in Section 6.1.1 is simply an orthogonal signalling system, with Hadamard functions for the basis set, turned into a binary code. Therefore when transmitted over an AWGN channel as described in the last section it is orthogonal signalling all over again. Thus methods for analysing its coding performance have already been given (Section 5.2.4), although this time the emphasis is on particular codes with fairly small block-sizes rather than on what happens in the limit of large codewords. (An exact

142 ERROR-CORRECTING CODES

method for calculating the error-probabilities is given in Appendix A. The results are applicable to any biorthogonal code on an AWGN channel.)

We plot the error-probability against γ_b, defined by (6.13). The error-probability chosen is the 'word-error' probability P_W, or the probability that a codeword is incorrectly decoded. Figure. 6.5 shows such plots for a number of codes, mostly the $(2^{k-1}, k)$ Reed–Muller codes:

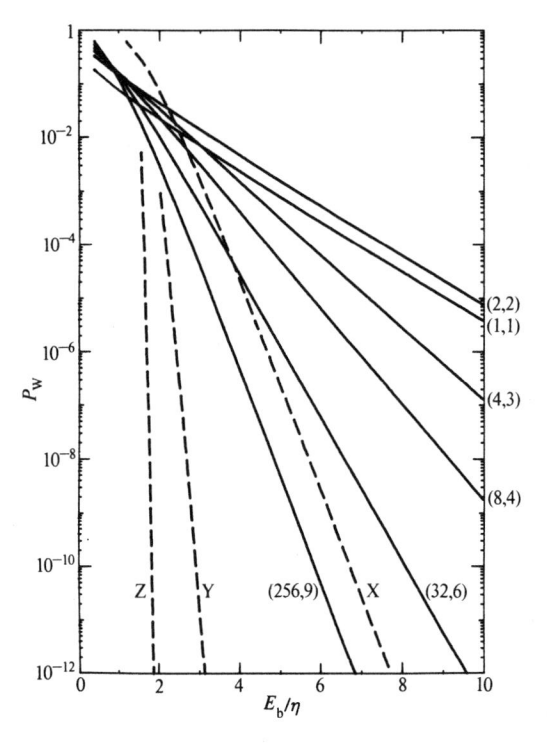

FIG. 6.5 Word-error probabilities for various codes with soft-decision decoding on the AWGN channel. The solid curves correspond to the $(2^{k-1}, k)$ Reed–Muller codes with $k = 1, 2, 3, 4, 6$ and 9. The dashed curves correspond to 'concatenated' codes to be described in Section 7.8.1.

The (1, 1) code is simply the raw binary transmission. Its error-probability is given by (6.11) and (6.12) as

$$P_e = \tfrac{1}{2} \operatorname{erfc}(\sqrt{\gamma_b}).$$

The (2, 2) code acts as though the bits in the raw binary transmission were paired off into 2-bit blocks, so that an error occurs if either bit (or both) is incorrectly determined. Thus the error-probability for this code is approximately twice that for the (1, 1) code. The (4, 3) code is a single-parity-check code.

As k is increased the codes give steeper and steeper plots. It can be shown (Section 5.2.4) that as $k \to \infty$ the plots have an asymptote determined by the Shannon bound, the vertical line at $\gamma_b = \ln 2 = 0.6931$. Unfortunately we note that the rate $k/n = k/2^{k-1}$ has become very low when $k = 9$. It turns out that the convergence to the asymptote is painfully slow. Rather large values of k are needed to do significantly better, and so we find that the values of $n = 2^{k-1}$ rapidly become impractical. Thus the (256, 9) code gives a performance close to the best that we can hope for from such codes. (The slowness of the convergence to the ideal limit is also discussed in Section 5.2.4 from another point of view.)

Other coding systems can do better. To encourage the reader to persevere the performance of some examples of so-called 'concatenated' codes called X, Y and Z in Fig. 6.5 are illustrated. These codes are described in Section 7.8. In spite of their large codewords they are fairly easily decoded, and so they give a practical way of achieving very low error probabilities ($P_W \approx 10^{-20}$) for quite reasonable values of γ_b.

The x-axis is usually set out logarithmically (Schwartz 1980, Section 6.4). A linear scale illustrates clearly how the error probability varies roughly exponentially with γ_b. (The plots would be straight lines if the behaviour were truly exponential. Actually they tend to steepen slightly as they fall.)

Often the 'bit-error' probability P_b is plotted rather than P_W. This is the probability that a randomly chosen bit in the decoded output is in error. The reason is that plotting P_W biases one against codes with larger values of k, since a single bit in error in the decoded word (of length k) counts as an error just as much as if every bit were wrong. None the less P_W has been chosen since this is the error-probability considered in the discussions on biorthogonal codes and on Shannon's theorem. In the case of the biorthogonal codes P_b is roughly one-half of P_W (provided k is not too small). This is because a decoding error usually means that the decoded word is selected at random since all the other codewords except one are at the same distance from any given codeword. Hence there is a 50 per cent chance that a given bit in the incorrectly decoded word will agree with the corresponding bit in the correct message-block.

6.4.3 Weakness of hard-decision decoding

We now compare the performances of some simple codes using hard- and soft-decision decoding on an AWGN channel. The factor $(t + 1)$ in (6.17) is (by (6.15)) not much larger than $\tfrac{1}{2}d_{\min}$, so that if this result is compared with (6.22), the corresponding result derived in the next section for the asymptotic behaviour of the error-probability of soft-decision decoding, it is found that the error probability is as though just over half the power had been used. In effect the hard-decision process has cost nearly 3 dB. This means in practice that for simple codes most of the advantage of coding is lost (Schwartz 1980, Section 6.8). Four systems over an AWGN channel with noise-spectral density η

are considered; in each case the word-error probability is found as a function of $\gamma_b = E_b/\eta$, where E_b is the energy available for each bit of information.

System A is simple uncoded four-bit transmission. Only hard decisions are possible at the receiver. The word error-probability is

$$P_A = 1 - (1 - p_A)^4, \quad \text{with} \quad p_A = \tfrac{1}{2} \operatorname{erfc}(\sqrt{\gamma_b}).$$

Here p_A is the bit-error rate given by (6.11), with the rate $R = k/n$ equal to 1, and P_A is computed as 1 minus the probability of finding all four bits incorrect.

System B uses minimum-Euclidean-distance decoding for the (8, 4) Reed-Muller code. The error probability is P_W as given by (A.2) in Appendix A with $k = 4$ in (A.1).

System C uses hard-decision decoding on the (8, 4) Reed-Muller code. A decoding failure or decoding error always occurs if there are two bit-errors or more, so that

$$1 - P_C = (1 - p_C)^8 + 8 p_C (1 - p_C)^7$$

with

$$p_C = \tfrac{1}{2} \operatorname{erfc}\{\sqrt{(\tfrac{1}{2}\gamma_b)}\}.$$

This code has $d_{\min} = 4$ and $t = 1$. In the general discussion we can only say that trouble might occur for more than t bit-errors. For this particular code it always does. The error-probability p_C for each bit is considerably increased above p_A since the total energy has to be shared among eight bits of the codeword rather than four.

Since most decoding failures are caused by double bit-errors which at least are detected it may seem unfair to compare this system with the other two systems where there is no provision for any kind of warning. So we introduce System D which uses the (7, 4) single-error correcting Hamming code described in Section 6.3.4. A false correction is made if there is a double error. The error probability P_D satisfies

$$1 - P_D = (1 - p_D)^7 + 7 p_D (1 - p_D)^6$$

with

$$p_D = \tfrac{1}{2} \operatorname{erfc}\{\sqrt{(4\gamma_b/7)}\}.$$

These probabilities are plotted in Fig. 6.6. We note how hard-limiting causes simple codes to lose nearly all their coding gain.

It must be noted that all this means is that hard-decision algorithms should not be used 'raw'. They may be very effectively used as the core of a soft-decision algorithm (Section 6.4.5).

BINARY BLOCK CODES ON THE AWGN CHANNEL 145

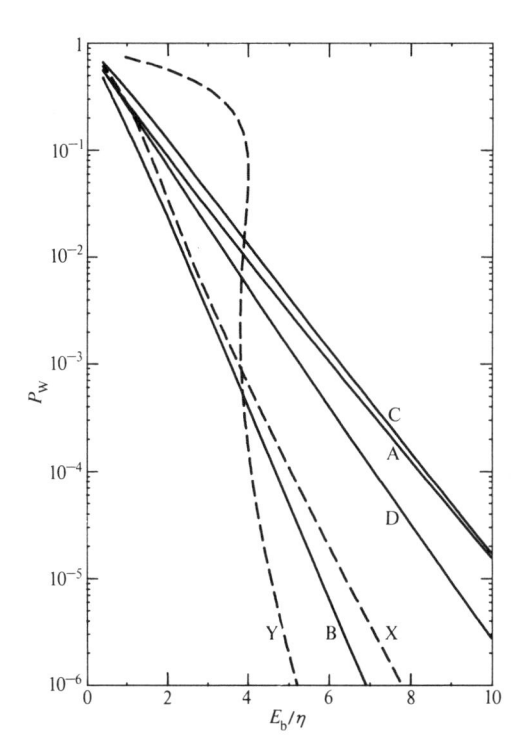

FIG. 6.6 Word-error probabilities for various codes. (A) (4, 4) uncoded; (B) (8, 4) Reed-Muller code with soft-decision decoding; (C) (8, 4) Reed-Muller code with hard-decision decoding; (D) (7, 4) Hamming code with hard-decision decoding; (X) (8, 4) Reed-Muller code used with hard-decision decoding with single-error correction on an ideal feedback channel; (Y) (8, 4) Reed-Muller code used with hard-decision error-detection on feedback channel.

6.4.4 *Hard-decision decoding with ARQ*

It is worth digressing briefly to see what kind of performance may be achieved with a simple code on an AWGN channel with feedback. Hard-decision decoding is used. The return channel over which requests for repeats (queries) are sent is ideal, so that such requests are sent back faithfully, instantaneously and with negligible consumption of energy or of any other resource.

Three things can happen at the decoder when a given word is received:

(a) There can be an error, incorrect decoding without warning. The probability of this happening is dented by p_x.

(b) There can be a decoding failure with a query. The probability for this is denoted by $p_?$.

(c) There can be correct decoding.

It can be shown that a particular codeword is sent on average $(1-p_?)^{-1}$ times, so that the energy per bit must be scaled up by this factor. The final error-probability P_e is also increased by this factor so that

$$P_e = (1-p_?)^{-1} p_x. \tag{6.23}$$

Suppose that the (8, 4) first-order Reed–Muller code (Table 6.2) is being used with hard decision followed by a decoder that corrects single errors and detects double errors. Suppose moreover that the all-zeros codeword $\mathbf{0}$ is sent. Let p be the probability that a codeword-bit is wrongly received. Then a query occur for all twenty-eight weight-2 error-patterns, for all fifty-six weight-4 error-patterns that are not codewords, and for all twenty-eight weight-6 error-patterns. An error occurs for all fifty-six weight-3 error-patterns, for all fifty-six weight-5 error-patterns, and for all eight weight-7 error-patterns, which are wrongly decoded, and for the weight-4 and weight-8 error-patterns which are codewords. Thus we find

$$p_x = 56p^3(1-p)^5 + 56p^5(1-p)^3 + 8p^7(1-p) + p^8, \tag{6.24}$$

and a similar result for $p_?$. In terms of the energy E per bit of codeword we have by (4.52) that

$$p = \tfrac{1}{2}\operatorname{erfc}(\sqrt{\gamma}) \tag{6.25}$$

with $\gamma = E/\eta$, and we have

$$\gamma_b = 2(1-p_?)^{-1}\gamma \tag{6.26}$$

for the energy per bit of information, scaled to take account of retransmissions. Curve X in Fig. 6.6 shows the performance of such a system. It is not markedly inferior to system B, soft-decision decoding without feedback. (The plot was made using γ as the independent variable.) Curve Y in Fig. 6.6 shows what happens if the same code is used with hard-decision error-detection only. An error occurs only if the error-pattern is a weight-4 codeword or the weight-8 codeword. All other error-patterns of weight greater than 0 cause a query. For $\gamma_b > 4$ this system performs better than any other! The basic reason for the difference between curves X and Y in Fig. 6.6 is that the error-correcting system X is fooled by weight-3 error-patterns while the error-detecting system in Y is fooled by (some) weight-4 error-patterns, which are far less likely when γ_b is large.

Soft-decision decoding with ARQ is also possible. For instance, one may choose a Euclidean distance a as a parameter. If the received point in signal-space is further than a from every signal-point then a query is issued. Otherwise minimum-distance decoding is employed.

6.4.5 Chase's soft-decision algorithm

An algorithm for soft-decision decoding proposed by Chase (1972) is now described. It comes close to the performance of minimum-distance decoding.

(Since minimum-distance decoding is not error-free it seems unnecessary to go to enormous lengths to get very close to its performance. This is like using very high-precision arithmetic in a formula which is inherently not exact.)

Suppose that polar signalling is used on an AWGN channel and the demodulator output u is transformed according to the formula

$$x = \tfrac{1}{2}(1 - u/\sqrt{E})$$

so that the receiver outputs $u = +\sqrt{E}$ and $u = -\sqrt{E}$ become respectively $x = 0$ and $x = 1$. Thus we have some received word $(x_1, x_2, \ldots, x_n)$ and we wish to find which is the nearest codeword $(c_1, c_2, \ldots, c_n)$ for which the square of the Euclidean distance $d_E^2 = \Sigma_i(x_i - c_i)^2$ is the least for all codewords. The total number of codewords is 2^k, and except on low rate codes like the Reed–Muller code trying each one in turn is not practicable.

We find the set of $l = \mathrm{int}(\tfrac{1}{2}d)$ received-word elements x_i which are nearest the value $\tfrac{1}{2}$. Here d is the Hamming distance of the code. These elements are the 'weakest' or most 'unreliable'. Next we make a list of 2^l codewords, not necessarily all distinct, produced by trying all 2^l possibilities for these weakest elements fed as bits into a hard-decision algorithm, together with the fixed rounded values of the other bits. Then the codeword nearest in Euclidean distance to the received word is chosen from this list. Evidently this algorithm uses a hard-decision algorithm repeatedly. For instance with the $d = 7$ Golay code the hard-decision algorithm, based on a syndrome-lookup table say, is used $2^3 = 8$ times. This should be very much faster than examining all $2^{12} = 4096$ codewords.

The effectiveness of this algorithm has been tested by computer simulation and found to be almost as good as true minimum-distance decoding.

6.4.6 *Viterbi decoding of block codes*

An algorithm for the Viterbi decoding of linear block codes was given by Wolf (1978). This is a true minimum-distance decoding algorithm, practicable for high-rate (n, k) codes where the number $r = n - k$ of check-sums is not large, say up to 10. The idea is as follows: Let $\mathbf{c} = (c_1, c_2, \ldots, c_n)$ be a codeword in a code with a parity-check matrix h_{ij}, $i = 1, 2, \ldots, r$, $j = 1, 2, \ldots, n$. We now consider the partial check-sums (for $i = 1, 2, \ldots, r$)

$$s_i(l) = \Sigma_{j=1}^{l} h_{ij} c_j \bmod 2 \qquad (l = 1, 2, \ldots, n), \quad s_i(0) = 0.$$

It is convenient to write these as r-bit strings:

$$\mathbf{s}(l) = \Sigma_{j=1}^{l} \mathbf{h}_j c_j, \qquad \mathbf{s}(0) = \mathbf{0},$$

where $\mathbf{h}_j$ is the jth column of h_{ij} and $\mathbf{s}(l)$ is the string $(s_1(l), s_2(l), \ldots, s_r(l))$. Naturally we must have

$$s(n) = 0$$

since the overall check-sums vanish (6.18).

We now set up a trellis (Section 6.2.1) where l (from 0 to n) labels the stages, and the 2^r possibilities for $s(l)$ the states. (There are of course this number of r-bit strings.) For example if there are two check-sums (i.e. $r = 2$), then s is a 2-bit string with four possibilities (00), (01), (10), and (11). Thus at each stage there are four states. We go from the $(l-1)$th stage to the lth stage by adding in an extra bit, so that

$$s(l) = s(l-1) + \mathbf{h}_l c_l.$$

This equation is the recipe for how to go from the state labelled by the string $s(l-1)$ at the $(l-1)$th stage to the state labelled by $s(l)$ at the lth stage. There are thus two possible transitions from any given state, since there are two possibilities for c_l. These transitions are represented by directed lines. We find that every possible codeword is represented by a path through this trellis starting in the state labelled by **0** at the stage $l = 0$ and finishing in the same state at the final stage $l = n$. Conversely any such path must correspond to some codeword, since its start and finish guarantee that the parity-check equations are obeyed.

Thereafter the decoding process is the same as for any other trellis code. For each choice of bit at the $(l-1)$th stage there is a 'cost' determined by the distance of the chosen bit from the real value that was actually received. Provided that the overall cost to be minimized is the sum of the costs associated with each transition we may use the usual Viterbi technique to find the minimum-cost path and hence the codeword nearest the received word.

The limitation on this algorithm is that the number of states must not be too large, so that $r = 10$ (giving 1024 states) seems about the best we can hope for. It should also be noted that the method is readily generalized to cases where the codeword elements are not bits, but symbols from a larger alphabet.

For further reading: Bhargava's article (1983) is a simple tutorial. There are several good books on coding theory, usually at a fairly advanced level. Here is a selection; Blahut (1983), Blake and Mullin (1975), Clark and Cain (1981), Lin and Costello (1983), McEliece (1977), MacWilliams and Sloane (1977), Wiggert (1978).

Exercises

6.1. Write down the codewords of the (4, 3) first-order Reed–Muller code and verify that it is a single-parity-check code.

6.2. Find the number of codewords A_d of weight d for the (8, 4) array code of Section 6.1.1 and the (9, 4) extended version, where there is an extra parity check on the parity-checks.

6.3. Find the minimum-cost path from S_{00} to S_{06} in the trellis of Fig. 6.2.

6.4. Prove by induction that the first-order $(2^{m-1}, m)$ Reed–Muller code is linear.

6.5. Show that adjacent codewords in an (n, n) Gray code have a Hamming distance of 1 (Section 3.2.3). The code is defined by the algorithm illustrated in Fig. 3.2(b), generalized to n-bit words.

6.6. Show that the minimum distance of the $(2^{m-1}, m)$ first-order Reed–Muller code is $\frac{1}{2} 2^{m-1}$. (Use induction.)

6.7. The Slepian standard array for an (n, k) binary linear code is a rectangular array of n-bit words with 2^k columns. It is constructed as follows: The top row of the array contains all the codewords c_i, with the codeword **0** in the left-hand or 'leading' position. Then of all the n-bit words which have not appeared so far in the array a word **u** of minimum weight is chosen. Then under each codeword c_i is placed the word $\mathbf{u} + c_i$. Note that **u** appears in the left-hand location, and hence it is called the leader of its row. Then the array is built up row-by-row in the same way, using as leader for each row a word of minimum weight which has not so far appeared anywhere in the array. The process stops when there are no more unused words. Show that (a) every n-bit word must appear at least once in the array, (b) no word appears more than once, (c) there are 2^{n-k} rows, (d) all the words in a row give the same syndrome, (e) words from different rows give different syndromes, (f) every syndrome corresponds to some row, and (g) the leader of any row is among the words of minimum weight in that row.

6.8. Show $P_e = 10^{-5}$ for raw transission when $\gamma_b = 9.6$ dB.

6.9. Show that on a BSC with $p = \frac{1}{2}$ every codeword is as likely as any other, with a given received word. Also show that from a BSC with $p > \frac{1}{2}$ we may obtain a BSC with $p < \frac{1}{2}$ if we put in an inverter which changes 0s into 1s and vice-versa.

6.10. Show that for a t-error correcting code

$$2^{n-k} \geqslant 1 + \binom{n}{1} + \binom{n}{2} + \ldots + \binom{n}{t}.$$

Show that the $t = 1$ Hamming codes obey this condition with equality, as does the triple-error correcting (23, 12) Golay code. Show also that a minimum-distance decoding algorithm for a code obeying this condition with equality is always fooled by $t + 1$ errors.

6.11. Assuming that there are no more than two errors, use the majority-logic decoding algorithm of Section 6.3.5 to correct errors in the corrupted codeword (101 010 110 011 100) of the (15, 7) code.

6.12. Suppose that a signal s_0 using the $(n, n-1)$ single-parity-check code is sent by polar pulse-modulation over an AWGN channel and the received word r

is so badly affected by noise that the nearest signal point s_1 differs in four places from s_0. Show that there is a signal point s_2 differing in two places from s_0 which is nearer to r than s_0 is. (Hint: Use contradiction.)

6.13. Derive (6.20). (Hint: Use the result at the end of Section 5.1.8.)

6.14. Extend the result of Exercise 6.12 to show that when the union bound or tangential union bound is being used to compute the error-probability for a single-parity-check code on an AWGN channel only signal points differing in two places from the transmitted signal need be considered. Find expressions for the union bound and the tangential union bound on the word-error probability of the $(n, n-1)$ single-parity-check code.

6.15. It can be shown (Section 5.2.4) that as $k \to \infty$ the plots in Fig. 6.5 have as asymptote the vertical line at $\gamma_b = \ln 2 = 0.6931$. Show that this result comes from Shannon's formula (5.41) and (5.42) in the limit of large bandwidth equivalent B.

6.16. Show that P_b is roughly one-half of P_W for a biorthogonal code (provided k is not too small). (Hint: A decoding error usually means that the decoded word is selected at random since all the other codewords except one are at the same distance from any given codeword.)

6.17. How would the (1, 1) and (2, 2) codes of Section 6.4.2 compare in a plot of P_b?

6.18. Find the asymptotic slopes of $\ln P_W$ against γ_b for large γ_b for the $(2^{k-1}, k)$ first-order Reed–Muller codes (Fig. 6.5).

6.19. Find the asymptotic slopes of the curves of $\ln P_W$ against γ_b for large γ_b for the cases, A, B, C, and D in Fig. 6.6.

6.20. Show that on the feedback channel of Section 6.4.4 a particular codeword is sent on average $(1-p_?)^{-1}$ times, and that the energy per bit must be scaled up by this factor. Also show that the final error-probability P_e is also increased by this factor to $(1-p_?)^{-1}p_x$.

6.21. Consider the (8, 4) code of Section 6.4.4 used as a single-error correcting code in the manner there described. Write down the result for $p_?$ akin to (6.24). Also find the expressions for p_x and $p_?$ when this code is used for hard-decision error-detection only.

7. Codes based on fields and polynomials

So far the only manipulations carried out with codewords have not been much more than the addition of binary strings bit by bit. Now the subject of algebraic coding theory is introduced, in which the symbols in a codeword are taken to be members of algebraic structures called finite fields. This makes possible the operations of multiplication and division, as well as addition and subtraction, so that completely new possibilities arise. The subject is a large one, and so the choice has been made to concentrate on those parts of the theory relevant to the theme of easily decodeable codes operating up to the Shannon limit. None the less the opportunity is taken to introduce codes (such as the BCH codes) which come naturally from the theory, although not directly relevant to the theme, and which have great practical and theoretical importance.

Finite fields are introduced in Section 7.1, and used to discuss BCH codes (Section 7.2). Polynomials over finite fields are discussed in Section 7.3 and are used in Section 7.4 to describe the family of cyclic codes. Then in Section 7.5 Reed–Solomon codes are introduced as parallels to polynomial interpolation in numerical analysis. Their close relationship to BCH codes is then demonstrated. These codes are directly relevant to the main theme. In Sections 7.6 and 7.7 the theory of algebraic decoding using the so-called Euclidean algorithm is developed; the complexity of this algorithm increases only as a low power of the codeword length rather than exponentially, and so it is vital to our purpose. Finally in Section 7.8 concatenated codes are introduced, which use double encoding, first by a Reed–Solomon code with algebraic decoding and then by a code with soft-decision decoding. Such codes combine the advantages of both types of decoding, and are indeed easily decodeable codes which can be made to approach the Shannon limit arbitrarily closely.

The mathematics in this chapter is different from that used previously, and the reader may be put off by its curious combination of elementary school algebra with a certain degree of sophistication. However the techniques are not hard, and the reader will be introduced to some of the most powerful codes known.

7.1 Fields

7.1.1 *Modular arithmetic*

We start by examining ways in which symbols in codewords may be used as quantities in a proper algebraic system, with multiplication and division as well as addition. We cannot use ordinary integers as symbols, since there is then no upper bound on the possible length, and if we try to impose a limit on the word-length, say 16 bits, then there is a risk of overflow. Moreover exact division is not possible, since ratios of integers are not integers in general.

This problem is overcome by the use of 'modular arithmetic', where we choose a positive integer p and keep only the remainders after integer division by p. It is helpful to introduce some notation at this stage:

(a) For p a positive integer and k any integer we denote by k mod p ('k modulo p') the integer in the range 0 to $p - 1$ obtained by adding to k or subtracting from k the appropriate multiple of p. (For k non-negative k mod p is the remainder after integer division.) This quantity will also be referred to as 'k reduced mod p'.

(b) The operation of multiplying two numbers and reducing the answer mod p will be called 'multiplication mod p'. Similarly for addition and subtraction.

Next we take the integers $0, 1, \ldots, (p - 1)$ and perform addition, subtraction, and multiplication on them mod p. This gives 'modular arithmetic', with only a finite set of quantities and no risk of overflow. Exact division will be introduced in due course. It is readily shown that the elementary rules of manipulation still work, and we have a simple example of an algebraic structure called a *ring* (in this case a 'commutative ring with an identity').

7.1.2 *Rings*

We now examine the rules of manipulation in a ring. First we list the simple rules for manipulating equalities. For any quantities or expressions a, b, c we have

(a) $a = a$;
(b) if $a = b$ then $b = a$;
(c) if $a = b$ and $b = c$ then $a = c$;
(d) a rule of substitution: if $a = b$ then we may replace any occurrence of a in an equality by b. Other rules can be derived from this one, including the previous two. Thus to show that $a + c = b + c$ follows from $a = b$ we start with $a + c = a + c$ and replace a by b on the right-hand side. Similarly $ac = bc$ follows from $a = b$.

A *ring* is a set of quantities or *elements* which can be added and multiplied together. The rules for a commutative ring with identity are:

(a) For any pair of elements in the ring there is a sum $a + b$ and a product ab (also denoted by $a.b$ or $a \times b$). The answer must also be an element in the ring. This is the simple but important *closure* property, and it guarantees that overflow need not occur in a finite ring.

(b) The operations of addition and multiplication are subject to *commutative*, *associative*, and *distributive* laws: For all a, b, c in the ring

$$a + b = b + a, \quad ab = ba;$$
$$a + (b + c) = (a + b) + c; \quad a(bc) = (ab)c;$$
$$a(b + c) = ab + ac.$$

The associative rule makes it unnecessary to bracket off a multiple product like $abcd$.

(c) There are two special elements in the ring, usually called 0 and 1, although they are not logically quite the same as the integers 0 and 1. They have the properties that for all a in the ring

$$a + 0 = a, \quad 1a = a.$$

It is easy to see that these elements are *unique*: for suppose there is another element $0'$ such that $a + 0' = a$ for all a. Then it follows that $0 = 0' + 0 = 0 + 0' = 0'$. Similarly for 1.

(d) For every element a in the ring there is a negative $-a$ also in the ring (closure!) with the property $a + (-a) = 0$. The sum $a + (-b)$ is usually written as $a - b$. Hence we have negatives and subtraction. It is easy to show that $-a$ is unique, for suppose both b and c give 0 when added to a. Then

$$b = b + 0 = b + (a + c) = (b + a) + c = 0 + c = c.$$

Elementary results then follow:

$$-(-a) = a, \quad (-a)b = -(ab), \quad (-a)(-b) = ab, \quad 0 = -0, \quad 0a = 0,$$

and $a + b = c$ implies $a = c - b$, as is shown by adding $-b$ to both sides.

It is readily shown that the modular arithmetic described above produces such a ring, in which p quantities $0, 1, \ldots, (p - 1)$ are added and multiplied mod p. This ring is denoted by Z_p. The addition and multiplication tables for Z_5 are shown in Table 7.1. The quantities 0 and 1 play their proper roles and the negative of any quantity is found from the addition table. Thus we see that $2 + 3 = 0$, so that $-2 = 3$.

7.1.3 *Fields*

So far we have not mentioned division or inverses. If we now put in an extra rule that every quantity except 0 has an inverse in the ring (closure again), then the ring becomes a *field*. The inverse of a is written a^{-1} and is defined by $a.a^{-1} = 1$, i.e. as that element which when multiplied by a gives 1. It is readily shown to be unique. (With b and c both inverses of a, evaluate bac two ways to show $b = c$.) This new rule will be called the *law of inverses*. Some consequences are $1^{-1} = 1, (a^{-1})^{-1} = a, (ab)^{-1} = a^{-1}b^{-1}$. We may write a/b for ab^{-1}

TABLE 7.1

+	0	1	2	3	4
0	0	1	2	3	4
1	1	2	3	4	0
2	2	3	4	0	1
3	3	4	0	1	2
4	4	0	1	2	3

·	0	1	2	3	4
0	0	0	0	0	0
1	0	1	2	3	4
2	0	2	4	1	3
3	0	3	1	4	2
4	0	4	3	2	1

From this rule follows the *law of cancellation*:

$ab = 0$ means either $a = 0$ or $b = 0$ or both.

(If $a \neq 0$ then multiply both sides of $ab = 0$ by a^{-1} to get $b = 0$.) This law is so called because if $ab = ac$ with $a \neq 0$ then $a(b - c) = 0$, and hence by the law $b - c = 0$, whence $b = c$.

The ring Z_5 is a field since every element except 0 has an inverse. Thus 2.3 = 1, whence $2^{-1} = 3$ and $3^{-1} = 2$. It can be shown that if p is a prime then Z_p is a field, otherwise not. (Thus in Z_{15} we see that 3.5 = 0 without either 3 = 0 or 5 = 0. Thus the cancellation law fails, and hence so does the law of inverses. However, some quantities in Z_{15} do have inverses. Thus we find 7.13 = 1.) The simplest field is obtained with $p = 2$, a prime value. This is the *binary field*. Its additions and multiplication tables are shown in Table 7.2. Other types of finite field are described in the next section. The best known infinite fields are the set of all integer ratios or rationals, the set of all real numbers (the 'real field'), and the set of all complex numbers.

Notation

(a) Usually we denote field-elements by lower-case Latin characters (with or without subscripts) but elements with special properties will be denoted by lower-case Greek letters such as α and β. The characters i to q will be reserved for ordinary integers, and z for a special element introduced in connection with polynomials (Section 7.3.1).

(b) For any positive integer n the notation a^n means the product of n factors a. We set $a^0 = 1$, and $a^{-n} = (a^{-1})^n$. The exponent n is an ordinary integer, never a field-element.

(c) A field with a finite number of elements is called a finite field or Galois field. This number is called the *order* of the field, and it is always some power of a prime number (including the prime number 2). There is only one field for each such order q, denoted by $GF(q)$. The fields with q prime are the rings

TABLE 7.2

+	0	1		·	0	1
0	0	1		0	0	0
1	1	0		1	0	1

Z_q and are called prime fields. The fields $GF(p^m)$ for p prime and $m > 1$ may be constructed as *extensions* of $GF(p)$, in a way to be outlined below. The extensions $GF(2^m)$ of the binary field are much the most widely used fields in coding theory, since their arithmetic is the easiest to implement on digital hardware, even easier than ordinary integer arithmetic.

Primitive elements. There is a very useful theorem peculiar to finite fields: If a finite field has order q, then it contains at least one primitive element α, say, with the property that all the powers $\alpha, \alpha^2, \alpha^3, \ldots, \alpha^{q-1}$ are distinct, and that $\alpha^{q-1} = 1 = \alpha^0$ (MacWilliams and Sloane 1977). Hence every non-zero field-element is equal to one of these powers. (Thus in Z_{11} the element 2 is primitive, with successive powers 2, 4, 8, 5, 10, 9, 7, 3, 6, 1.)

Parallels. Some parallels and differences with the familiar real field are these: Linear algebra concerned with linear equations over a finite field (that is, with coefficients and unknowns in the finite field) is much the same as over the real field. Thus linear equations can be solved by Gaussian elimination, determinants are defined as usual, and matrices can be inverted if the determinant does not vanish. In fact the solution of linear equations on a computer is simplified by the absence of rounding errors and of overflow. Polynomials can be manipulated as described below, much as for the real field. Polynomial equations cannot always be solved, keeping within the field, and some elements (like -1 in the real field) may not have square roots. On the other hand it is not in general possible to define a scalar product (u, v) of vectors with elements in a finite field so that $(u, u) \neq 0$ if $u \neq 0$. Thus algebraic theorems that depend on being able to normalize non-zero vectors may not work. There is no concept of ordering or size in a finite field, nor of limiting processes, so no calculus or transcendental functions. Finally a finite field contains primitive elements, unlike the real field.

7.1.4 *Extensions and* $GF(2^m)$

Any field $GF(p)$ may be extended into $GF(p^m)$ for any positive integer m in a manner analogous to the way that the real field is extended into the complex field. The simplest but logically least satisfactory way of doing the latter is simply to 'invent' a new element j with the property that $j^2 = -1$. Of course this process needs to be justified. (This may be done by showing that manipulations in the new complex field are disguised manipulations of real numbers.) Thereafter the closure property requires that all elements of the form $a + jb$ (with a and b in the old real field) are included, as are powers of j greater than 1, although these may be 'boiled down' using the rule $j^2 = -1$.

We may analogously extend $GF(2)$ into $GF(2^m)$. We take $m = 4$ for illustration. We invent a new element α with the property $\alpha^4 = \alpha + 1$, which, like j in complex-number theory, should be regarded as a 'known' constant. (How and why this works will be explained in due course.) Once we add α to our field the closure property automatically requires that polynomial expressions like $\alpha + 1$, $\alpha^6 + \alpha^5 + 1$, and so on also belong to this field. It must be noted that, for any element a in a binary extension field $GF(2^m)$, $a + a = (1 + 1)a = 0a = 0$, so that $-a = a$. In fact minus is the same as plus! (None the less the minus sign will continue to be used whenever it seems natural, or whenever it would reappear in a generalization to non-binary fields.) Powers of α greater than or equal to 4 may be boiled down by using the rule $\alpha^4 = \alpha + 1$. Thus we end up with a field of 16 elements of the form

$$c_0 + c_1\alpha + c_2\alpha^2 + c_3\alpha^3, \tag{7.1}$$

where the c_i and 0 or 1. After we have multiplied any two such field elements together we reduce powers of α greater than or equal to 4 by using the rule $\alpha^4 = \alpha + 1$. We may represent an expression like (7.1) in a computer simply by specifying the coefficients as components of a 4-bit string (c_0, c_1, c_2, c_3). In this way we may add, (subtract), multiply, and divide these strings purely by manipulations in the binary field. It turns out that α is a primitive element because its successive powers give all the non-zero elements of the field, as shown in Table 7.3.

In evaluating α^4 we have used the defining equation for the first time, and in evaluating α^8 we have used $\alpha + \alpha = (1 + 1)\alpha = 0\alpha = 0$. Although α is a primitive element, α^3 is not since its fifth power ($=\alpha^{15}$) is 1.

For typographical convenience field-quantities will ususally be represented as in the left-hand column of Table 7.3. The product (ratio) of two quantities is given by adding (subtracting) the exponents mod 15. Addition is not quite so simple. We may add the corresponding bit-strings in the right-hand column. Thus $\alpha^{11} + \alpha^{13} \to (0111) + (1011) = (1100) \to \alpha^4$. (A quicker method for adding two field-elements uses 'Zech's logarithms' defined by $1 + \alpha^p = \alpha^{Z(p)}$ for $p = 1, \ldots, 14$. Thus $Z(5) = 10$ since $1 + \alpha^5 = \alpha^{10}$. Then

$$\alpha^p + \alpha^q = \alpha^p(1 + \alpha^{q-p}) = \alpha^{p+Z(q-p)}.$$

The function $Z(p)$ is tabulated in Table 7.4. Using this table we find $\alpha^{11} + \alpha^{13} = \alpha^{11+Z(13-11)} = \alpha^{11+8} = \alpha^4$.)

Subfields. A set of elements from a field that form a field in their own right is called a subfield. Thus the set $\{0, 1, \alpha\}$ is not a subfield of $GF(16)$

FIELDS 157

TABLE 7.3
*Successive powers of α expressed in the form
(7.1) and represented as 4-bit strings.*

α^0	1	(1000)
α^1	α	(0100)
α^2	α^2	(0010)
α^3	α^3	(0001)
α^4	$\alpha + 1$	(1100)
α^5	$\alpha^2 + \alpha$	(0110)
α^6	$\alpha^3 + \alpha^2$	(0011)
α^7	$\alpha^3 + \alpha + 1$	(1101)
α^8	$\alpha^2 + 1$	(1010)
α^9	$\alpha^3 + \alpha$	(0101)
α^{10}	$\alpha^2 + \alpha + 1$	(1110)
α^{11}	$\alpha^3 + \alpha^2 + \alpha$	(0111)
α^{12}	$\alpha^3 + \alpha^2 + \alpha + 1$	(1111)
α^{13}	$\alpha^3 + \alpha^2 + 1$	(1011)
α^{14}	$\alpha^3 + 1$	(1001)
α^{15}	1	(1000)

TABLE 7.4
Zech logs Z(p) defined by $1 + \alpha^p = \alpha^{Z(p)}$

p	Z(p)	p	Z(p)
1	4	8	2
2	8	9	7
3	14	10	5
4	1	11	12
5	10	12	11
6	13	13	6
7	9	14	3

since $\alpha + 1$ and $\alpha\alpha = \alpha^2$ do not belong to this set, and hence the closure property is not satisfied. The set $\{0, 1\}$ is a subfield of $GF(16)$, as is the set $\{0, 1, \alpha^2 + \alpha, \alpha^2 + \alpha + 1\}$. The order of any subfield of $GF(p^m)$ is $p^{m'}$ with m' a factor of m.

7.2 Hamming and BCH codes

7.2.1 *Hamming codes*

As a useful application of the above theory we consider the structure of certain codes.

Binary Hamming codes (Section 6.3.4) are single-error correcting codes with codewords of $n = 2^m - 1$ bits, and their structure can be based on $GF(2^m)$. Let us use $GF(16)$. We choose a primitive element α and require that the codeword bits c_i satisfy the 'check-sum condition'

$$\sum_{i=0}^{14} c_i \alpha^i = 0, \qquad (7.2)$$

where the c_i belong to the binary subfield, and α is a member of the main field. (Note that the bits are numbered starting at 0.) We then write

$$\alpha^i = \sum_{j=0}^{3} h_{ji} \alpha^j \qquad (i = 0, 1, \ldots, 14) \qquad (7.3)$$

(with h_{ji} in $GF(2)$), as in the second column of Table 7.3. The matrix h_{ji} is shown in Table 7.5, with the columns given by the right-hand entries in Table 7.3. Then by (7.3) we obtain the binary check-sum conditions

$$\sum_{i=0}^{14} h_{ji} c_i = 0, \qquad j = 0, 1, 2, 3 \qquad (7.4)$$

by equating to zero the coefficients of each power of α from 0 to 3. Thus h_{ji} is the parity-check matrix for this code (Exercise 7.9). We see that c_0 is expressed as a linear combination of $c_4, \ldots, c_{14}$ by the first equation $c_0 = c_4 + c_7 + c_8 + c_{10} + c_{12} + c_{13} + c_{14}$ and similar statements can be made about c_1, c_2, and c_3. Thus the 11 bits c_4 to c_{14} can be arbitrarily chosen to convey information and hence they are called 'information symbols'. Thereafter c_0 to c_3 are determined as 'parity checks'. We have a (15, 11) code.

To see how an error is corrected let us suppose that the received bits are

$$r_i = c_i + e_i \qquad (7.5)$$

where only one of the e_i, e_p, say, is non-zero. Then we see that by (7.2)

$$\sum_i r_i \alpha^i = \sum_i e_i \alpha^i = \alpha^p.$$

This expression can be represented as a 4-bit string and p is found by seeing which power of α matches.

Example. Correct the error in (110 110 101 110 101), assuming that there is not more than one.

TABLE 7.5

i \ j	0	1	2	3	4	5	6	7	8	9	10	11	12	13	14
0	1	0	0	0	1	0	0	1	1	0	1	0	1	1	1
1	0	1	0	0	1	1	0	1	0	1	1	1	1	0	0
2	0	0	1	0	0	1	1	0	1	0	1	1	1	1	0
3	0	0	0	1	0	0	1	1	0	1	0	1	1	1	1

Using the right-hand column of Table 7.3 we find that

$$\Sigma r_i \alpha^i \to (1000) + (0100) + (0001) + \ldots + (1001) \to \alpha^9.$$

Thus the bit in location 9 must be changed, to give (110 110 101 010 101).

7.2.2 BCH codes

Bose-Ray-Chaudhuri-Hocquenghem (BCH) codes correct two or more errors (MacWilliams and Sloane 1977). The idea will be illustrated by the (15, 7) double-error correcting code based on $GF(16)$. (This is in fact the same as the (15, 7) code of Section 6.3.5.) As well as (7.2) we impose the extra check-sum conditions

$$\Sigma_{i=0}^{14} c_i \alpha^{2i} = 0, \qquad (7.6)$$

$$\Sigma_{i=0}^{14} c_i \alpha^{3i} = 0. \qquad (7.7)$$

The first equation in fact imposes no extra constraints since

$$(\Sigma_i c_i \alpha^i)^2 = \Sigma_i c_i^2 \alpha^{2i} = \Sigma_i c_i \alpha^{2i},$$

and so (7.6) is implied by (7.2). Note that on squaring a sum of terms in a binary extension field the cross-terms add up in pairs to zero, and that the quantities c_i in $GF(2)$ satisfy

$$c_i^2 = c_i. \qquad (7.8)$$

However (7.7) imposes four new binary check-sum conditions like (7.4). Hence there are eight parity-check bits and only seven information bits, and so the code is a (15, 7) code.

Suppose we have errors at locations p and q. To find these we first compute the *syndromes* (in $GF(16)$)

$$s_1 = \Sigma_i r_i \alpha^i = \alpha^p + \alpha^q = x + y,$$

$$s_3 = \Sigma_i r_i \alpha^{3i} = \alpha^{3p} + \alpha^{3q} = x^3 + y^3,$$

where r_i is given by (7.5) and where

$$x = \alpha^p, \qquad y = \alpha^q. \qquad (7.9)$$

Hence we find

$$s_1^3 = x^3 + y^3 + xy(x + y) = s_3 + xys_1.$$

(Note how 3 is replaced by 1 in a binary extension field!) Thus we know $x + y = s_1$ and $xy = (s_1^3 + s_3)/s_1$, so that x satisfies the quadratic $x(s_1 + x) + (s_1^3 + s_3)/s_1 = 0$. The unknown y also satisfies this quadratic. We then solve for x and y, usually by trying all 16 cases to see which makes the left-hand side vanish. (The usual method for solving quadratics by formula does not work in $GF(2^m)$, since 2 becomes 0.) There are some special cases: $s_1 = s_3 = 0$ means that there are apparently no errors. There is one error if $s_1^3 + s_3 = 0$ without $s_1 = 0$. There is a decoding failure due to three or more errors if $s_1 = 0$ without $s_3 = 0$, or if the quadratic has no solutions (in the main field $GF(16)$). Once we have found x and y we locate the error by finding the exponents p and q in (7.9).

Example: Verify that (100 010 111 000 000) is a codeword. The check-sums (syndromes) are $s_1 = \alpha^0 + \alpha^4 + \alpha^6 + \alpha^7 + \alpha^8$ and $s_3 = \alpha^0 + \alpha^{12} + \alpha^3 + \alpha^6 + \alpha^9$ (using $\alpha^{15} = 1$), both of which vanish.

Example: Find the errors in the received word (111 010 111 000 000) assuming that there are not more than two.

The check-sums are $(0110) \to \alpha^5$ and $(0010) \to \alpha^2$. The quadratic is then $x^2 + \alpha^5 x + \alpha^3 = 0$. Trial and error shows that the solutions are $x = \alpha^1$ and α^2, so that the errors are in locations 1 and 2.

More generally to set up a binary BCH code we choose a field $GF(2^m)$ for some m. Each codeword consists of $n = 2^m - 1$ bits $c_0, \ldots, c_{n-1}$, which ae made to satisfy the check-sum conditions

$$\sum_{i=1}^{n-1} c_i \alpha^{il} = 0, \qquad l = 1, 2, \ldots, r \tag{7.10}$$

for some r. Here α is a primitive element in $GF(2^m)$ and the c_i belong to the binary subfield. Conversely an n-bit word is a codeword if the bits satisfy (7.10).

It can be shown that t errors can be corrected if $r \geq 2t$. The equations in (7.10) for even l are redundant, although this is only true with the c_i restricted to the binary subfield.

7.3 Polynomials

7.3.1 *Polynomials and their degrees*

We now continue the development of the algebraic theory by introducing polynomials.

An extra element z is introduced into $GF(q)$ which shares the properties of a ring with the field elements, so that there is no guarantee of an inverse for any algebraic expression containing z. All the powers of z are distinct, z^0 is 1, and negative powers are not defined. An expression of the form

$$A(z) = a_m z^m + a_{m-1} z^{m-1} + \ldots + a_1 z + a_0$$

is called a polynomial *over* the field $GF(q)$ to which the coefficients $a_m, a_{m-1}, \ldots, a_0$ belong. Polynomials will be denoted by capital letters. The degree of this polynomial, denoted by $\deg(A)$ or $\deg A$, is the highest power of z with a non-zero coefficient, m in this case if $a_m \neq 0$. If $a_m \neq 0$ then a_m is called the leading coefficient and $a_m z^m$ the leading term. Two polynomials are equal if their corresponding coefficients are equal, any unmatched coefficients being zero. Thus the equality $A(z) = 0$ asserts that all the coefficients of $A(z)$ are zero. (Polynomial equalities in this sense are often referred to as *identities*.) The sum of two polynomials is given by adding the corresponding coefficients, and the product is defined by using the distributive law and collecing together the terms with the same power of z. The use of z is in fact just a mnemonic device because the sum and product of the two polynomials can be defined purely in terms of the coefficients. It is easy to show that the cancellation law does apply to polynomials. A non-zero field-element or *scalar* can be regarded as a polynomial of degree zero, and 0 is sometimes regarded as a polynomial of degree $-\infty$. Thus polynomials of degree zero do have inverses. The rules for degrees are:

deg (AB) = deg (A) + deg (B),
deg $(A + B)$ = maximum of deg (A) and deg (B), if these are not equal,
deg $(A + B) \leqslant$ deg (A) if deg (A) = deg (B),

since it is then possible that the leading terms cancel.

The expression

$$A(c) = a_m c^m + a_{m-1} c^{m-1} + \ldots + a_1 c + a_0$$

with c a field-element will be called a *polynomial expression*, since it is a field-element or scalar, and not a polynomial, unless we care to regard it as a polynomial of degree zero. Replacing z by c in a polynomial equality gives an equality in the field. Thus if $C(z) = A(z)B(z)$ then $C(c) = A(c)B(c)$.

If $A(c) = 0$, then c is called a *root* of the polynomial $A(z)$.

7.3.2 *Factors and multiples*

We say that P divides Q or that P is a *factor* of Q if there is a polynomial R satisfying $Q = PR$. We call Q a *multiple* of P, and we sometimes write $R = Q/P$. If R is a scalar we say that Q is a *scalar multiple* of P. If A is a factor of B and B is a factor of C then it immediately follows that A is a factor of C. If A is a factor of B and of C then by this result and by the distributive law it is also a factor of $LB + MC$ where L and M are any polynomials. A polynomial of degree $\geqslant 1$ is *irreducible* if it has no factors except scalars and scalar multiples of itself. All first-degree polynomials are irreducible. The unique factorization theorem states that polynomials may be factored into irreducible factors essentially in only one way (except for the order of the factors and the transfer of scalar multiples between factors). A proof is given in Appendix B.

7.3.3 *Division algorithm*

We may divide a polynomial G by a polynomial F to give a *quotient* Q and *remainder* R:

$$G = QF + R \quad \text{with} \quad \deg(R) < \deg(F).$$

(If $\deg(G) < \deg(F)$ then $Q = 0$ and $R = G$; if F is a scalar then $R = 0$ and $Q = F^{-1}G$.) This is done by repeatedly reducing the degree of what is left of G by subtracting $F(z)$ multiplied by a suitable power of z times a scalar. The calculation is usually laid out as a long division. The result is unique since if there are other polynomials Q' and R' such that $G = Q'F + R'$ with $\deg R' < \deg F$, then subtraction gives $(Q - Q')F = (R' - R)$ with the degree of the LHS greater than that of the RHS.

Example: Table 7.6 shows the division of $G(z) = \alpha z^4 + \alpha^5 z$ by $F(z) = z^2 + \alpha z + \alpha^2$ (as polynomials over $GF(16)$ so that subtraction is the same as addition).

TABLE 7.6

$$
\begin{array}{r}
\alpha z^2 + \alpha^2 z \\
z^2 + \alpha z + \alpha^2 \overline{\big)\,\alpha z^4 + \alpha^5 z} \\
\alpha z^4 + \alpha^2 z^3 + \alpha^3 z^2 \\
\hline
\alpha^2 z^3 + \alpha^3 z^2 + \alpha^5 z \\
\alpha^2 z^3 + \alpha^3 z^2 + \alpha^4 z \\
\hline
\alpha^8 z.
\end{array}
$$

Here we first subtract $\alpha^2 z^2 F(z)$ from $G(z)$ to leave a 'partial remainder' of degree less than $\deg G$ and then we subtract $\alpha^2 z F(z)$ from that. The process cannot be repeated since the degree of the new partial remainder is less than $\deg F$. Thus the quotient is $\alpha z^2 + \alpha^2 z$ and the remainder $\alpha^8 z$.

7.3.4 *The mod operation and congruences*

We define $G \bmod F$ as the remainder R when G is divided by F. It is readily seen that

$$(AB) \bmod F = (A \bmod F)(B \bmod F) \bmod F,$$

and

$$(A + B) \bmod F = (A \bmod F) + (B \bmod F).$$

The mod operation can be performed before or after adding or multiplying, depending on whichever is more convenient. The *congruence* $A \equiv B \bmod F$ means that $A \bmod F = B \bmod F$, or equivalently that $(A - B)$ is a multiple of

F. Congruences may be manipulated like equalities as described in Section 7.1.2 (provided of course that they are all modulo the same polynomial). In particular the law of substitution is readily shown to be valid. However, cancellation is not usually allowed (Section 7.6.3). A congruence is usually weaker than an equality, but if it is known that $\deg(A - B) < \deg(F)$, then the equality $A = B$ follows from $A \equiv B \bmod F$.

7.3.5 *Polynomial rings*

Just as we constructed a ring Z_p by choosing a positive integer p and then adding and multiplying the quantities $0, \ldots, (p-1) \bmod p$, so also it is possible to construct a ring by choosing a polynomial $G(z)$ of degree m over $GF(q)$ and then carrying out the addition and multiplication of polynomials over $GF(q)$ of degree less than m by first adding and multiplying in the usual way and then reducing the result mod G. The number of polynomials in the ring is q^m, the number of polynomials of degree less than m. If moreover $G(z)$ is irreducible, then the ring becomes a field, since an inverse can then be found for every non-zero polynomial in the ring. (See Section 7.6.2.) Extension fields may be constructed in this way. In particular the binary extension field $GF(2^4)$ described in Section 7.1.4 is constructed by adding and multiplying mod $z^4 + z + 1$ polynomials over $GF(2)$ of degree less than 4. The element 0 is the null polynomial, the element 1 is the polynomial $1 + 0z + 0z^2 + 0z^3$, and the mysterious element α is the polynomial z or $0 + 1z + 0z^2 + 0z^3$. Thus the product α^4 or $z.z.z.z$ is z^4 reduced mod $z^4 + z + 1$, which is (over $GF(2)$) the polynomial $z + 1$, i.e. $\alpha + 1$. (Note that with multiplication redefined in the manner just described $z.z.z.z$ is not the same as the polynomial z^4.)

7.4 Cyclic codes

7.4.1 *Generator polynomial*

One way of obtaining an n-symbol codeword $(c_0, c_1, \ldots, c_{n-1})$ from a k-symbol message $(b_0, b_1, \ldots, b_{k-1})$ is to obtain the c_i as the coefficients of the polynomial $C(z) = \Sigma_i c_i z^i$ produced by multiplying $B(z) = \Sigma_i b_i z^i$ by a fixed *generator* polynomial $J(z)$ of degree $n-k$. It is assumed that the alphabet contains q symbols with q a prime power, so that each symbol can be represented by an element of $GF(q)$; usually the symbols are bits from $GF(2)$. Thus the polynomial $C(z)$ is a multiple of $J(z)$. It will be assumed that J is not a multiple of z, since then so is C and hence c_0 is always 0 and conveys no information. The code is not systematic since the message symbols do not appear explicitly in the codeword. A modification of the encoding technique soon remedies this problem. The message polynomial $B(z)$ is multiplied by z^{n-k} to leave $(n-k)$ zero coefficients clear for accepting the check-sums. Then we divide by $J(z)$,

$$B(z)z^{n-k} = Q(z)J(z) + R(z), \qquad \deg(R) < n-k.$$

so that $B(z)z^{n-k} - R(z)$ is a multiple of $J(z)$. This is then our codeword polynomial with the information symbols as coefficients of the higher powers of z, and the check-sums, the coefficients of $-R$, at the lower end.

This division algorithm is easy to implement, especially for polynomials over $GF(2)$. For example, consider the (8, 3) code with $J(z) = z^5 + z^3 + 1$ and let $B(z)$ be $z^2 + z + 1$, corresponding to the message $(b_2 b_1 b_0) = (111)$ with $k = 3$ bits. At this stage the bits will be numbered starting with 0 on the right. Then the division of $z^5 B(z)$ by $J(z)$ can be written out as follows, using the notation of 'detached coefficients' where the coefficients are shown without the powers of z. These powers decrease from left to right:

```
                  110
          ┌─────────────
   101001 │ 11100000
            101001
          ─────────────
             10001
             101001
          ─────────────
             10110
```

Thus we find that $Q(z) = z^2 + z$, $R(z) = z^4 + z^2 + z$, and the codeword is (111.10110), with the dot showing where the remainder is tagged on to the original message. The division process is usually carried out using a feedback shift-register (Figure 7.1a), holding $n - k$ (=5) bits. The register is first loaded with the first $n - k$ bits of the dividend, the remaining bits waiting on the right. Then the bits are shifted left and the left-most bit, as well as being placed in the output where the quotient is built up, is *fed back* into the shift-register, into which it is added at the indicated places *after* the shift. This process is carried out k times (Figure 7.1b, c, d), until all the dividend has been entered. The remainder is left in the register. The feedback connections are made at the cells corresponding to the non-zero coefficients in $J(z) - z^{n-k}$, if the cells are numbered starting with 0 on the right.

The ease with which this process can be implemented makes this type of code very popular for error-detection on a binary channel. The encoder computes the remainder produced by dividing the message-polynomial by $J(z)$ and tags it on to the end of the message. This makes a codeword, a multiple of $J(z)$. The remainder is known as a *cyclic redundancy check* (CRC) in this context. At the receiving end the received word is divided by $J(z)$, this division proceeding as the received word comes in, so that there need be very little delay caused by the processing. If the final contents of the feedback shift-register (that is, the remainder) are all zeros, then the received word is a codeword

CYCLIC CODES

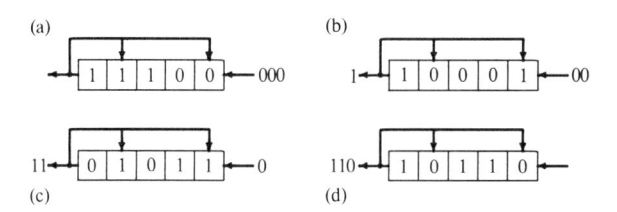

(a)   (b)

| 1 | 1 | 1 | 0 | 0 |←000   1←| 1 | 0 | 0 | 0 | 1 |←00

11←| 0 | 1 | 0 | 1 | 1 |←0   110←| 1 | 0 | 1 | 1 | 0 |←
(c)   (d)

FIG. 7.1. Feedback shift-register for implementing division of $(z^2 + z + 1)z^5$ by $z^5 + z^3 + 1$

and it is presumed that no error has occurred. But otherwise an error is detected, since the received word is not a multiple of $J(z)$.

The CCITT recommendation V41 uses this system of error-detection with the choice

$$J(z) = z^{16} + z^{12} + z^5 + 1$$

(Bylanski and Ingram 1980, CCITT Green Book VIII. CCITT is the abbreviation for Comité Consultative International de Telephonie et de Telegraphie, an international committee set up to promote telecommunications standards. See Graham 1983.)

7.4.2 Cyclic property

It can be shown that there is a positive integer N for which $J(z)$ is a factor of $z^N - 1$ (provided J is not a multiple of z). Hence $z^N - 1$ could be a codeword polynomial giving a codeword of the low Hamming weight of 2. Hence the code would have a Hamming distance of 2, no better than a single-parity-check code. Thus we restrict the codeword length n by $n \leq N$ to keep the degree below N. If we choose $n = N$, then $J(z)$ is a factor of $z^n - 1$ and end-around or cyclic shifts of codewords are also codewords! Hence, the name *cyclic code*. For if $C(z) = c_{n-1}z^{n-1} + \ldots + c_0$ is a codeword (i.e. a multiple of J. of degree less than n) then $zC(z)$ is also a codeword being a multiple of J, unless $c_{n-1} \neq 0$, when its degree is n and thus too high. The end-around shift is equivalent to replacing the term $c_{n-1}z^n$ by c_{n-1} or to subtracting $c_{n-1}(z^n - 1)$. The shifted word $zC(z) - c_{n-1}(z^n - 1)$ is then a codeword since the degree is less than n, and since it is a multiple of J, both terms being multiples of J. Finding n for given J is done by long division of $z^n - 1$, that is of $z^n + 1$, if we are working over $GF(2)$. Since n is not known we do not yet know where to put the 1. So we start off with z^n and keep dividing until the result of a subtraction leaves a solitary 1, and then we put the 1 above that. The method is illustrated in Table 7.7, with $J(z) = z^4 + z^2 + z + 1$. Only the coefficients are shown (in order of decreasing powers of z).

TABLE 7.7

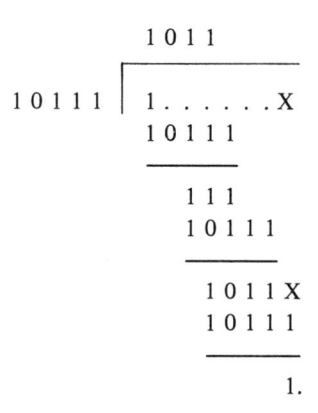

```
                1 0 1 1
        ┌─────────────────
1 0 1 1 1 │ 1 . . . . . . X
          1 0 1 1 1
          ─────────
            1 1 1
            1 0 1 1 1
            ─────────
              1 0 1 1 X
              1 0 1 1 1
              ─────────
                  1 .
```

If the bit X is 0 then the remainder is as shown but if X is 1 then the remainder is 0. Thus we find that $z^7 + 1$ gives a remainder 0 so that $n = 7$ in this case. The maximum value of n is $2^r - 1$, where $r = \deg(J) = n - k$, basically because that is the greatest possible number of different partial remainders in the long-division process for z^n, before the solitary 1 appears.

Example: The (15, 11) Hamming code. Every codeword polynomial satisfies $C(\alpha) = 0$ by (7.2). (This polynomial expression and the others in this example all have coefficients in the subfield $\{0, 1\}$.) Now α satisfies $A(\alpha) = 0$ where $A(z) = z^4 + z + 1$. By the division algorithm we can find R of degree less than 4, such that $C = QA + R$ and hence that $R(\alpha) = C(\alpha) - Q(\alpha)A(\alpha) = 0$. This is only possible with $\deg R < 4$ if $R(z) = 0$. Hence C is a multiple of A, which is thus the generator.

Example: The (15, 7) BCH code. By (7.2), (7.6), and (7.7) the codewords satisfy $C(\alpha) = 0$, $C(\alpha^2) = 0$, and $C(\alpha^3) = 0$. The generator is thus the lowest degree polynomial over $GF(2)$ with roots α, α^2, α^3. This turns out to be the lowest common multiple of $A(z)$ and the similar polynomial for α^3, $B(z) = z^4 + z^3 + z^2 + z + 1$, in fact $AB = z^8 + z^7 + z^6 + z^4 + 1$. (The element α^2 is also a root of $A(z)$.) Hence (100 010 111 000 000) and all its cyclic shifts are codewords in the (15, 7) code.

Example: The Golay code. The generator is

$$J(z) = z^{11} + z^{10} + z^6 + z^5 + z^4 + z^2 + 1,$$

a factor of $z^{23} + 1$ (over $GF(2)$).

Even the Reed–Muller codes are closely related to cyclic codes.

7.5 Reed-Solomon Codes

Reed-Solomon codes are based on a slightly different use of polynomials. There is only one large field $GF(q)$ this time. Messages and codeword symbols belong to this field. If $q = 2^m$ each symbol can be represented by an m-bit string, as in the right-hand column of Table 7.3.

Suppose we are given a table of values of a polynomial $R(z)$ over the real-number field. The polynomial is known to be of second degree. The values are: $R(0) = 0$, $R(1) = 1$, $R(2) = 4$, $R(3) = 9$, $R(4) = 16$. To find the polynomial we need to find 3 coefficients, and it is evidently $1z^2 + 0z + 0$. Thus two entries in the table are redundant. But if any two entries are erased we can still find the polynomial. Suppose now that instead there is a misprint and the fourth entry reads $R(3) = 8$. We are told that there is one error. We can find it by plotting a graph or by techniques from numerical analysis. To correct the error we need to know both where it is and how big it is. This needs two pieces of information which are provided by the double redundancy. So we can find the error and correct it. More generally we might expect to be able to correct a table with s erasures and t errors provided that the number r of redundant entries satisfies $r \geq s + 2t$. At any rate this is what can be done by algebraic decoding. The number r equals $n - k$, where k is the number of polynomial coefficients (the degree of the polynomial plus 1), and n the number of entries in the table.

7.5.1 *The code*

In a Reed-Solomon code a finite field $GF(q)$ is used instead of the field of real numbers, and message and codeword symbols are elements of this field. If the message is $(b_0, b_1, \ldots, b_{k-1})$ we set up

$$R(z) = b_{k-1}z^{k-1} + \ldots + b_1 z + b_0 \tag{7.11}$$

of degree $k - 1$ and evaluate it for n ($\geq k$) replacements of z by distinct field elements $a_0, a_1, \ldots, a_{n-1}$ to be specified in a moment. (Thus $n \leq q$.) The codeword is then the string $(c_0, c_1, \ldots, c_{n-1})$ with $c_i = R(a_i)$. It is thus a table of values of a polynomial of known degree and the decoder has to find the polynomial in spite of errors and erasures. If there are only erasures, then difference methods from numerical analysis may be used to interpolate the polynomial and find the coefficients (Burden, Faires, and Reynolds 1981). The *redundancy* is defined by $r = n - k$.

Although other possibilities do exist, it will be assumed from now on that all the non-zero field elements in $GF(q)$ are used to form the set $\{a_i\}$, so that $n = q - 1$. Hence without loss of generality we may set $a_i = \alpha^i$ ($i = 0, 1, \ldots, n - 1$) where α is a primitive element in $GF(q)$, satisfying $\alpha^n = 1$.

Example: The (15, 11) Reed-Solomon code over $GF(16)$. Here the message string is $k = 11$ symbols long. Let the message be $(\alpha, 0, 0, 0, 0, 0, 0, 0, 0, 0, \alpha^2)$, so that $R(z) = \alpha + \alpha^2 z^{11}$, a polynomial over $GF(16)$. Then

$$c_0 = R(\alpha^0) = \alpha + \alpha^2 = \alpha^5,$$
$$c_1 = R(\alpha^2) = \alpha + \alpha^{12} = \alpha^{13},$$
$$c_2 = R(\alpha^2) = \alpha + \alpha^{22} = \alpha + \alpha^7 = \alpha^{14},$$
$$c_3 = R(\alpha^3) = \alpha + \alpha^{32} = \alpha + \alpha^2 = \alpha^5, \text{ etc.}$$

Thus the codeword is $(\alpha^5, \alpha^{13}, \alpha^{14}, \alpha^5, \alpha^{13}, \alpha^{14}, \ldots)$. Using the representation of Table 7.3 the codeword contains 60 bits, and starts as (0110 1011 1001 ...). The code is a (60, 44) binary code.

Example: 'Sharing a secret' (Shamir 1979). Suppose that the k coefficients of $R(z)$ form a secret message. Suppose moreover that the coded values $c_i = R(a_i)$ are handed out, one value to each of n people. Then the message is concealed until at least k of them get together to determine the coefficients of $R(z)$.

This form of encoding is not systematic. For this we choose $R(z)$ to satisfy $R(a_i) = b_i$ for $i = 0$ to $k - 1$, instead of setting its coefficients to the message symbols. Thus $R(z)$ is an interpolating polynomial. The entries $R(a_i)$ for the other values of a_i may be found by standard methods from numerical analysis.

7.5.2 Check-sum conditions

We now set about finding the check-sum conditions as the usual decoding algorithm is based on them. We first need a simple lemma. If $\alpha^n = 1$ then

$$\sum_{i=0}^{n-1} \alpha^{i(j-l)} = \begin{cases} \text{'}n\text{'} & \text{if } j - l \text{ is a multiple of } n, \\ 0 & \text{otherwise.} \end{cases} \quad (7.12)$$

Here 'n' is the field element obtained by adding n 1s, which is just 1 in a binary extension field, since $n = q - 1$ is odd. (Equation (7.12) is the analogue in a finite field of a discrete Fourier transform. See Exercise 2.20.) To prove this lemma we first set $b = \alpha^{j-l}$, so that $b^n = 1$. The sum is

$$s = 1 + (b + b^2 + \ldots + b^{n-1}) = b^n + (b + \ldots + b^{n-1}) = b(1 + b + \ldots + b^{n-1}) = bs.$$

Thus $(1 - b)s = 0$ and so either $b = 1$, when the sum is 'n', or else $s = 0$.

From (7.11) we find

$$c_i = R(\alpha^i) = \sum_{j=1}^{n-1} b_j \alpha^{ij}$$

with the extra coefficients set to zero:

$$b_{n-1} = b_{n-2} = \ldots = b_k = 0. \quad (7.13)$$

We define $b_n = b_0$, and note that $\alpha^n = 1$. Thus it follows that

REED-SOLOMON CODES

$$c_i = \sum_{j=0}^{n-1} b_j \alpha^{-i(n-j)}$$
$$= \sum_{j=1}^{n} b_j \alpha^{-i(n-j)} = \sum_{l=0}^{n-1} b_{n-l} \alpha^{-il} \qquad \text{for } i = 0, \ldots, n-1.$$

Thus using our lemma we find

$$\sum_{i=0}^{n-1} \alpha^{ij} c_i = \sum_{l=0}^{n-1} b_{n-l} \left[\sum_{i=0}^{n-1} \alpha^{i(j-l)} \right] = \text{'}n\text{'} \, b_{n-j},$$

since the factor in square brackets allows only the term with $l = j$ to survive (see (7.12)). Hence by (7.13) we obtain the check-sum conditions

$$\sum_i \alpha^{ij} c_i = 0 \qquad \text{for } j = 1, 2, \ldots, n-k. \tag{7.14}$$

Thus the Reed–Solomon codes are closely related to the BCH codes (see (7.10)). However, the code-symbols c_i are not restricted to a subfield. Over $GF(2^m)$ a codeword is $2^m - 1$ symbols long or $m(2^m - 1)$ bits long, if each symbol is represented by m bits.

7.6 Euclid's algorithm

7.6.1 *The algorithm*

Euclid's algorithm is very useful in the theory of fields, not only because it demonstrates that the inverses of the elements in the so-called fields described in Section 7.1.3 and Section 7.3.5 really do exist, but also because it enables one to compute these inverses easily. Our main interest in it is that it provides the basis of an algebraic 'Euclidean' decoding algorithm, suitable not only for Reed–Solomon codes but for other codes which arise naturally out of the discussion, such as the Goppa codes (Section 7.7.5). In its simplest form it provides a way of finding the highest common factor (h.c.f.) of two integers. The larger is divided by the lesser, and then repeatedly the old divisor is divided by the latest remainder until zero is obtained. The h.c.f. is the number obtained just before the zero. Thus starting with 392 and 273 we obtain the sequence 392, 273, 119, 35, 14, 7, 0, giving 7 as the h.c.f. of 392 and 273. The reason is that any common factor of the divisor and dividend is also a factor of the remainder. A similar process can be implemented for polynomials. In the following discussion the entities will be taken to be polynomials over any specified field, which may well be the field of real numbers. However, the integer parallel will be used for purposes of illustration.

A relation like

$$P_{i+1} = -A_i P_i + P_{i-1}, \qquad i = 1, 2, \ldots \tag{7.15}$$

is called a three-term recurrence relation. Any set of polynomials $Q_0, Q_1, \ldots$ satisfying

$$Q_{i+1} = -A_i Q_i + Q_{i-1}, \qquad i = 1, 2, \ldots \tag{7.16}$$

will be called a solution of (7.15). So long as the A_i are known it is evident that a solution is uniquely specified by Q_0 and Q_1. We need two lemmas. The first is that any linear combination of two solutions

$$R_i = LP_i + MQ_i$$

is also a solution. The proof is trivial. The second lemma will be called the 'cross-product' theorem. Define the 'cross-product'

$$X_i = P_i Q_{i+1} - Q_i P_{i+1} \quad \text{for } i = 0, 1, 2, \ldots \quad (7.17)$$

Then multiplying (7.15) by Q_i and (7.16) by P_i and subtracting eliminates the term containing A_i to leave $-X_i = X_{i-1}$, and so we have

$$X_i = (-1)^i X_0, \quad (7.18)$$

so that X_i is a constant except for sign.

We now combine these ideas with the use of the division algorithm. Given polynomials G and F we set $P_0 = G$, $P_1 = F$. Starting with $i = 1$ we choose A_i to be the quotient and P_{i+1} to be the remainder when P_{i-1} is divided by P_i, that is

$$P_{i-1} = A_i P_i + P_{i+1}, \text{ with } \deg P_{i+1} < \deg P_i. \quad (7.19)$$

The former of these relations is (7.15) rearranged. The process must stop with $P_{N+1} = 0$ for some N, either because the previous divisor P_N is a scalar or because more generally it is a factor of P_{N-1}. We write H for P_N because in fact it is the h.c.f. (highest degree common factor) of F and G. This method of finding the h.c.f. is called Euclid's algorithm. Using the same coefficients A_i we now set up several solutions Q_i, R_i, and Y_i of (7.15), as shown in Table 7.8 and as explained in the commentary to be read with it, rather than after it.

Commentary on Table 7.8: In lines 1 and 2 P_{i+1} is the remainder and A_i is the quotient when P_{i-1} is divded by P_i. Start with $P_0 = G$, $P_1 = F$, and stop at $i = N$ with $P_{N+1} = 0$. Set $H = P_N$.

In lines 3 and 4 start Q_i and R_i as shown. By the cross-product theorem ((7.17) and (7.18)) on Q_i and R_i

TABLE 7.8

i	0	1	...	N	$N+1$
A_i		A_1	...	A_N	
P_i	G	F	...	H	0
Q_i	0	1	...	Q_N	Q_{N+1}
R_i	1	0	...	R_N	R_{N+1}
Y_i	$-Q_N$	R_N	...	0	$(-1)^N$

$$Q_N R_{N+1} - R_N Q_{N+1} = (-1)^{N+1}. \quad (7.20)$$

In line 5 set Y_i as the linear combination $Y_i = R_N Q_i - Q_N R_i$. Hence the values of Y_0, Y_1, and Y_N. Equation (7.20) gives Y_{N+1}. (Note trivial special cases. If F is a factor of G then $N = 1$ and $H = F$, and if $F = 0$ then $N = 0$ and $H = G$.)

Example: The parallelism between integer division and polynomial division enables us to set up a version of Euclid's algorithm for integers. An example of this process is shown in Table 7.9, starting with $G = 25$ and $F = 9$. The values of the P_i are obtained as follows: 9 goes into 25 twice remainder 7. Next 7 goes into 9 once remainder 2, and so on. The A_i are the quotients. Here N is found to be 4.

It should be noted that for finding h.c.f.s only the P_i are needed. To find inverses (see next section) the A_i, P_i, and Q_i are needed. The last two lines are not really part of the algorithm, but are used in the theory.

7.6.2 Some consequences

An important consequence is the existence of an inverse 'mod G'. Other results are needed to justify the decoding algorithm.

(a) The result

$$\deg(Q_{i+1}) > \deg(Q_i), \quad i = 1, 2, \ldots \quad (7.21)$$

is easily proved by induction from (7.16), using the result $\deg(A_i) > 0$, which follows at once from (7.19).

(b) By means of the cross-product theorem on P, Q, on P, R, and on P, Y we find

$$G = H Q_{N+1} (-1)^N, \quad (7.22)$$

$$F = H R_{N+1} (-1)^{N+1}, \quad (7.23)$$

and

$$G R_N + F Q_N = H. \quad (7.24)$$

From (7.22) and (7.23) we see that H is a factor common to both F and G, and from (7.24) that any common factor of F and G is also a factor of H. Hence the name highest common factor for H.

(c) We call F and G *relatively prime* if there is no (non-scalar) factor common to both F and G. In this case H is a scalar with an inverse. (For if H is not a scalar then by (7.22) and (7.23) it is a common factor of F and G.) It follows from (7.24) that

$$GM + FL = 1, \quad (7.25)$$

172   CODES BASED ON FIELDS AND POLYNOMIALS

TABLE 7.9

i	0	1	2	3	4	5
A_i		2	1	3	2	
P_i	25	9	7	2	1	0
Q_i	0	1	−2	3	−11	25
R_i	1	0	1	−1	4	−9
Y_i	11	4	3	1	0	1

with

$$L = H^{-1}Q_N, \quad M = H^{-1}R_N. \tag{7.26}$$

Moreover we see that

$$\deg(L) = \deg(Q_N) < \deg(Q_{N+1}) = \deg(G) \tag{7.27}$$

by (7.26), (7.21), and (7.22). In the integer parallel shown in Table 7.9 we find that $L = -11$ and $M = 4$, so that (7.25) gives $4 \times 25 + 9 \times (-11) = 1$.

(d) From (7.25) we see that $FL = 1 \bmod G$, and from (7.27) that deg $(L) < \deg(G)$. We call L the *inverse* of F mod G, and denote it by $F^{-1} \bmod G$. It is unique, for if L' is another inverse such that $FL' \equiv 1 \bmod G$ and $\deg(L') < \deg(G)$ then by substituting $FL' \equiv 1 \bmod G$ in $L'FL$ we find $L'FL = L \bmod G$. Similarly $L'FL \equiv L' \bmod G$. Hence $L \equiv L' \bmod G$ so that $(L - L')$ is a multiple of G. Thus $L = L'$ since the degree of $L - L'$ is less than $\deg(G)$.

In the integer parallel the answer is not quite immediate. We find that it gives -11 as the inverse of 9 mod 25. To make this a positive value we augment it by 25, to give 14 as the inverse of 9 mod 25 in the range 0 to 24, i.e. 9×14 mod $25 = 1$.

(e) We now consider the ring of polynomials of degree less than G, with addition and multiplication mod G. (See Section 7.3.5.) From the above result we find that every polynomial in the ring which is relatively prime to G has a unique inverse in the ring. Moreover, Euclid's algorithm gives a practical way of checking that any polynomial is relatively prime to G and at the same time of finding the inverse if it exists.

(f) Now suppose that G is irreducible. Then every polynomial in the ring (except 0) is relatively prime to G and so has an inverse. The ring has now become a field. This is one way of constructing extension fields. The integer version of Euclid's algorithm provides a parallel result, that in integer arithmetic mod p every integer from 1 to $(p-1)$ has an inverse mod p, provided p is prime. This is why the methods of constructing fields described in Section 7.1.4 and Section 7.3.5 work.

7.6.3 Properties of inverses mod G

We extend our notation so that any inverses appearing in a congruence mod G are to be evaluated mod G.

(a) We have a weak cancellation law to the effect that we may cancel a factor F from both sides of a congruence 'mod G' provided F is relatively prime to G. This is because we may multiply both sides by F^{-1} mod G. It is easy to show the validity of using one congruence ($FF^{-1} \equiv 1$ in this case) to substitute for a term or expression in another, so that FF^{-1} may be replaced by 1.

(b) There is a rule similar to the rule for adding ordinary fractions: $a/b + c/d = (ad + bc)/(bd)$. Suppose that the polynomial S is given by the equality

$$S = AB^{-1} + CD^{-1} \qquad (7.28)$$

where B and D are relatively prime to G and where these inverses and the others to be used are taken 'mod G'. Then we find

$$S \equiv PQ^{-1} \bmod G \qquad (7.29)$$

where

$$P = AD + BC, \quad Q = BD. \qquad (7.30)$$

For Q is relatively prime to G so that Q^{-1} exists. Then from $Q^{-1}BD \equiv 1 \bmod G$ we obtain $Q^{-1} \equiv B^{-1}D^{-1}$ mod G by multiplying both sides by B^{-1} and D^{-1}. Then we substitute for P and Q^{-1} on the right of (7.29).

This result is readily extended to the sum of three or more fractions.

7.6.4 Examples of inverses

(a) The following equality will be needed in the next section:

$$(1 - az)^{-1} \bmod z^r = 1 + az + (az)^2 + \ldots + (az)^{r-1}. \qquad (7.31)$$

To verify this we note that multiplying the right-hand side by $(1 - az)$ gives $1 - (az)^r$, which reduced mod z^r gives 1. Moreover the degree is less than r.

(b) If $G(a) \neq 0$ then $(z - a)$ is relatively prime to $G(z)$ and

$$(z - a)^{-1} \bmod G(z) = -\{G(a)\}^{-1} [\{G(z) - G(a)\}/(z - a)]. \qquad (7.32)$$

Note that this is an equality, not a congruence, and that $G(a)$ is a scalar possessing an inverse. The second factor in square brackets is the polynomial obtained by removing the factor $z - a$ from $G(z) - G(a)$. We verify this result by multiplying on the right by $(z - a)$ to give 1 plus a multiple of $G(z)$. The degree of the right-hand side is one less than deg (G).

174   CODES BASED ON FIELDS AND POLYNOMIALS

7.7 Algebraic decoding

7.7.1 *Introductory*

The stage is now set for describing an algebraic decoding algorithm for an (n, k) Reed-Solomon code over $GF(q)$, capable of correcting both errors and erasures. A 'key equation' is constructed from the syndromes. This may be solved in a number of ways, one of which is based on the Euclidean algorithm (Sugiyama, Kasahara, Hirawawa, and Namekawa 1976). Another algorithm discovered by Berlekamp is easier to implement, but why it works is not so easy to understand. (See Blahut 1983 or Massey 1968 for a description.)

7.7.2 *The key equation*

We start with the check-sums (7.14) obeyed by any codeword $(c_0, \ldots, c_{n-1})$ of a Reed-Solomon code. (We recall that these c_i are elements in a large Galois field $GF(q)$, that α is a primitive element in this field, and that $n = q - 1$.) We set $l = j - 1$, multiply by z^l, and sum over l from 0 to $(r - 1)$, where $r = n - k$ is the redundancy. Then we employ (7.31) to obtain the polynomial equalities

$$0 = \sum_{l=0}^{r-1} z^l \sum_{i=0}^{n-1} \alpha^{il} \alpha^i c_i$$
$$= \sum_{i=0}^{n-1} c_i \alpha^i \sum_{l=0}^{r-1} (\alpha^i z)^l = \sum_{i=0}^{n-1} c_i \alpha^i (1 - \alpha^i z)^{-1}, \qquad (7.33)$$

where the inverses are mod z^r. After moving scalar factors around on the right-hand side we find

$$\sum_{i=0}^{n-1} c_i (z - \alpha^{-i})^{-1} = 0 \qquad \text{(inverses mod } z^r\text{)}. \qquad (7.34)$$

Let us suppose that in transmission errors e_i are added to the c_i to give a received word $(r_0, \ldots, r_{n-1})$ with $r_i = c_i + e_i$. (The e_i and r_i belong to the same field as the c_i.) The syndrome polynomial is defined by

$$S(z) = \sum_{i=0}^{n-1} r_i (z - \alpha^{-i})^{-1} \qquad \text{(inverses mod } z^r\text{)}. \qquad (7.35)$$

The degree is less than r. By reversing the steps in (7.34) and (7.33) we find

$$S(z) = -\sum_{i=1}^{r} s_i z^{i-1}$$

with

$$s_i = \sum_{j=0}^{n-1} r_j \alpha^{ij}.$$

Hence the coefficients of $S(z)$ are the negatives of the syndromes. (Of course the minus can be forgotten in a binary extension field.) From (7.14) it follows that

$$s_i = \sum_{j=0}^{n-1} e_j \alpha^{ij}.$$

ALGEBRAIC DECODING 175

Example: Consider the (15, 11) Reed-Solomon code over $GF(16)$ with errors in r_0 and r_1 given by $e_0 = \alpha^{10}$, $e_1 = \alpha^{12}$. (If the $GF(16)$ elements are represented by binary strings as in Table 7.3, then the error pattern is (1110, 1111, 0000, 0000, ... 0000).) The syndromes are $s_1 = \alpha^{10} + \alpha^{13} = \alpha^9$, $s_2 = \alpha^{10} + \alpha^{14} = \alpha^{11}$, $s_3 = \alpha^{10} + \alpha^{15} = \alpha^5$, and $s_4 = \alpha^{10} + \alpha^{16} = \alpha^8$. (The sums are conveniently calculated by Zech's logarithms.) Thus $S(z) = \alpha^8 z^3 + \alpha^5 z^2 + \alpha^{11} z + \alpha^9$.

An alternative expression is obtained for $S(z)$ by using (7.35) and (7.34). We find

$$S(z) = \Sigma_{i=0}^{n-1} (c_i + e_i)(z - \alpha^{-i})^{-1}$$
$$= \Sigma_{i=0}^{n-1} e_i (z - \alpha^{-i})^{-1} \quad \text{(inverses mod } z^r\text{)}.$$

Let us now suppose that there are t errors, that is t of the quantities e_i are non-zero. ($t = 2$ in the above example.) We rename these quantities as y_k for k from 1 to t, and write x_k for the corresponding values α^{-i}. Then we find

$$S(z) = \Sigma_{k=1}^{t} y_k (z - x_k)^{-1} \quad \text{(inverses mod } z^r\text{)}.$$

This is a sum of fractions, and by using the rule of addition (7.28) to (7.30) extended to t fractions we find

$$S \equiv PQ^{-1} \bmod G \tag{7.36}$$

where

$$G(z) = z^r, \tag{7.37}$$

$$P(z) = \Sigma_k y_k \Pi_{l(\neq k)} (z - x_l), \tag{7.38}$$

$$Q(z) = \Pi_k (z - x_k), \tag{7.39}$$

the sum and the products being taken from 1 to t. The polynomial Q is called the error-locator, since its roots determine the values of the α^{-i} at the error-locations. It has degree t exactly, whereas P has degree less than t. After multiplying (7.36) through by Q we obtain the *key equation* (actually a congruence)

$$P \equiv QS \bmod G. \tag{7.40}$$

There are $2t$ quantities x_k and y_k to be determined, and r coefficients in the polynomial $S(z)$ from which they must be found. Hence it is not surprising that we need the restriction $t \leq \frac{1}{2}r$. We require a solution of (7.40) satisfying $\deg P < \deg Q \leq r/2$. In general $QS \bmod G$ is of degree $r - 1$, and so we obtain linear equations for the coefficients of Q by demanding that the coefficients in $QS \bmod G$ of powers of z greater than or equal to $r/2$ must vanish. The problem can be solved this way, but a better method will be described in a moment. Once we have found Q we can find P by (7.40). The error locations are found by trying all the possible (big) field elements α^{-i} out as roots of $Q(z)$. To help in obtaining y_k it should be noted that if we replace z by x_k in (7.38) all terms vanish except the one containing y_k.

7.7.3 Euclidean algorithm

A quick way to solve the key equation (7.40) is as follows. If $P_{i-1} \equiv Q_{i-1}F$ mod G and $P_i \equiv Q_iF$ mod G, then the linear combinations given by (7.15) and (7.16) satisfy $P_{i+1} \equiv Q_{i+1} F$ mod G. We start as in Table 7.8 with the trivial solutions $P_0 = G$, $Q_0 = 0$, and $P_1 = F$, $Q_1 = 1$, where $F = S$. These solutions do not have the right degrees. However, iterating the Euclidean algorithm makes the degree of P_i fall steadily with i, and that of Q_i to rise. As soon as $\deg P_i < \deg Q_i$ we set $Q = cQ_i$, $P = cP_i$, where c is a scalar to make the leading coefficient of Q equal to 1. A justification for this procedure is given in Appendix B.

Example: Consider the error pattern discussed in the example in Section 7.7.2. The computation of Table 7.8 is shown in Table 7.10. (For typographical reasons the rows have been rewritten as columns.) We see that $\deg P_3 < \deg Q_3$ and so we find

$$Q(z) = z^2 + \alpha^3 z + \alpha^{14}, \qquad P(z) = \alpha^3 z + \alpha^8.$$

The roots of $Q(z)$ are readily found to be $x_1 = \alpha^0 (=1)$ and $x_2 = \alpha^{-1}$. To determine the values of the errors y_1 and y_2 at locations 0 and 1 we use

$$P(x_1) = y_1(x_1 - x_2) \quad \text{or} \quad P(\alpha^0) = y_1(\alpha^0 - \alpha^{-1}),$$

and

$$P(\alpha^{-1}) = y_2(\alpha^{-1} - \alpha^0).$$

The first equation gives

$$\alpha^{13} = y_1 \alpha^3, \text{ whence } y_1 = \alpha^{10}; \text{ similarly we obtain } y_2 = \alpha^{12}.$$

7.7.4 Erasures as well

Suppose there are s erasures and t errors, with

$$s + 2t \leq r. \tag{7.41}$$

TABLE 7.10

i	A_i	P_i	Q_i
0		z^4	0
1	$\alpha^7 z + \alpha^4$	$\alpha^8 z^3 + \alpha^5 z^2 + \alpha^{11} z + \alpha^9$	1
2	$\alpha^7 z + \alpha^2$	$\alpha z^2 + \alpha^4 z + \alpha^{13}$	$\alpha^7 z + \alpha^4$
3	$\alpha^{14} z + \alpha^{10}$	$\alpha^2 z + \alpha^7$	$\alpha^{14} z^2 + \alpha^2 z + \alpha^{13}$
4	$\alpha^3 z + \alpha^8$	α^{14}	$\alpha^{13} z^3 + \alpha^3 z^2 + \alpha^7 z + \alpha^5$
5		0	αz^4

The erasures are filled in arbitrarily and S is computed as before. Even a lucky guess will be treated as an error, although at the end the computed value of the error will be zero. There are then $s + t$ errors of which s are already located. The error locator polynomial may be broken into two factors R and Q of degrees s and t respectively, of which R, the product of factors $(z - \alpha^{-i})$ corresponding to the erasures, is already known. Moreover

$$\deg(P) \leqslant s + t - 1, \qquad \deg(Q) = t. \tag{7.42}$$

The key equation becomes

$$P \equiv Q(RS) \bmod G$$

with RS known and with

$$\deg(P) < s + \deg(Q).$$

We use the Euclidean algorithm as before, but starting with $F = (RS \bmod G)$, and proceeding until $\deg(P_i) < s + \deg(Q_i)$. Why the restriction (7.41) matters is explained in Appendix B.

7.7.5 Variations on a theme

The recasting of the check-sum equations for Reed–Solomon codes into the form (7.34) which was done for the sake of developing a decoding algorithm also points to a much larger family of codes which can also be decoded this way. We choose an arbitrary polynomial $G(z)$ over $GF(q)$ of degree r, a set of distinct quantities a_i from $GF(q)$, and another set of quantities b_i ($\neq 0$) also from $GF(q)$. Then we set up a code by requiring that the n quantities c_i from $GF(q)$ form a codeword if and only if

$$\sum_{i=0}^{n-1} c_i b_i (z - a_i)^{-1} = 0 \qquad \text{(inverses mod } G(z)\text{)}. \tag{7.43}$$

Since the quantities a_i must be distinct we see that $n \leqslant q$. Moreover the a_i must *not* be roots of $G(z)$, for if $G(a_i) = 0$ then $(z - a_i)$ is not relatively prime to $G(z)$ and the inverse mod $G(z)$ does not exist. The Reed–Solomon codes are a special case of such codes, with $G(z) = z^r$, $a_i = \alpha^{-i}$, and $b_i = 1$. It is evident that the Euclidean decoding algorithm can decode these more general codes just as readily as Reed–Solomon codes, since nowhere is $G(z) = z^r$ used.

It is also possible to form codes by requiring that the codeword-symbols c_i, as well as satisfying (7.43), lie in a subfield of $GF(q)$. Thus a binary BCH code has a check-sum equation (7.10) just like the one for Reed–Solomon codes (7.14), but the codeword symbols lie in the binary subfield. In general the Euclidean decoding algorithm cannot take account of this, except to report a decoding failure if the 'corrected' value of a symbol does not lie in the subfield.

However when the c_i are restricted to the binary subfield (in the case when the main field is a binary extension field) then it is obvious that we need only one piece of information to correct each error, namely its location. The value of the error can only be 1. Under these circumstances the use of the Euclidean algorithm is rather clumsy, and it is preferable to use the Berlekamp algorithm. (This algorithm is also good for Reed-Solomon codes. Its relationship to the Euclidean algorithm is briefly touched on in the appendix.)

The large family of codes defined through (7.43) contains many important codes such as the Goppa codes. (For further details see McEliece 1977.)

7.8 Concatenated codes

7.8.1 Double encoding

A typical Reed-Solomon code over $GF(2^m)$ has $n = 2^m - 1$ symbols per codeword, and each symbol may be represented by m bits. Hence the codeword is $m(2^m - 1)$ bits long. The input to the encoder is a block of k symbols ($k < n$), which may again be represented by km bits.

It is evident that such a code is not every effective on a channel with random errors because the loss of even one bit from a symbol means the loss of the whole symbol of m bits. This is because the decoding algorithm has no way of taking account of symbols that are 'nearly right'.

However, the code is much better if the errors or erasures come in bursts, since the loss of jm consecutive bits means the destruction of at most $(j + 1)$ symbols. The code is also very effective on certain quantum channels (see Chapter 8) where again whole symbols rather than individuals bits tend to be lost.

A Reed-Solomon code can be made much more effective on a channel with random errors (or erasures) if each m-bit symbol coming out of the encoder is then encoded again into a longer p-bit string by a (p, m) binary encoder. The method is shown in Fig. 7.2. A message-block of km bits is encoded by a Reed-Solomon encoder which treats the block as k symbols. The output is a code-word of n symbols which are individually put through (p, m) binary encoders. The final codeword is then np bits long, so that the overall rate is (k/n) (m/p). At the decoding end the process is reversed. As far as the Reed-Solomon decoder is concerned the effect of the second or inner coding/decoding process is to trade moderately frequent bit-errors for much rarer symbol errors, for when the inner decoder fails it usually returns something badly wrong, or it may even signal an erasure. In any case this is no worse than a single-bit error as far as the Reed-Solomon decoding algorithm is concerned. This double-encoding and -decoding process is shown in Fig. 7.3, and the overall (pn, km) code is called a *concatenated code* (Forney 1966). The inner code can be well matched to the channel with soft-decision decoding, and in this way we combine the advantage of this type of decoding with the power of algebraic decoding.

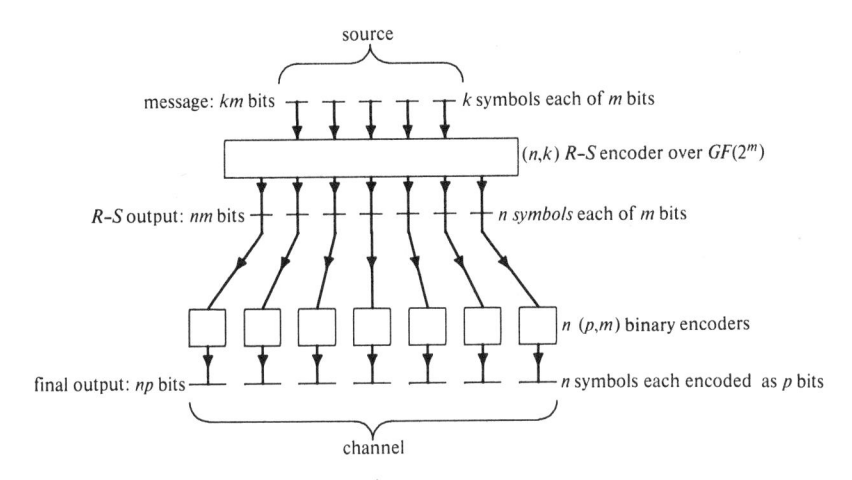

FIG. 7.2 Encoding a concatenated code.

FIG. 7.3 Arrangement of encoders and decoders in a concatenated scheme.

It should be mentioned in passing that the relation $n = 2^m - 1$ between the number of symbols n in a Reed-Solomon codeword and the number of bits per symbol m need not be adhered to. The number of symbols n' may be made less than $2^m - 1$. The simplest way is to modify Reed-Solomon codes by replacing the primitive element α by a non-primitive element β, which satisfies $\beta^{n'} = 1$, where n' is a submultiple of n. To go the other way we restrict the code-symbols to a subfield $GF(2^{m'})$ so that the number m' of bits per symbol is a submultiple of m, so that $n > 2^{m'}$.

Example: The symbols from the (15, 11) Reed-Solomon encoder (over $GF(16)$) are each represented by four bits. Suppose they are re-encoded by the (8, 4) Reed-Muller code described in Section 6.1.1. Thus each input block to the Reed-Solomon encoder consists of 11 symbols or 44 bits and the output is 16 symbols or 60 bits. The final output from the Reed-Muller encoder consists of 120 bits per block. Thus the code is a binary (120, 44) code. The curve X in Fig. 6.5 shows the error performance of this code. It is assumed that an error occurs if the Reed-Muller decoding returns more than two incorrect decodings in any block. The Reed-Muller decoding is taken to be the soft-decision process described in Section 6.1.3. The curves Y and Z in Fig. 6.5 are the error performances of more powerful codes of this sort, a (63, 43)

Reed–Solomon code over $GF(64)$ concatenated with the (32, 6) binary Reed–Muller code, and a (511, 451) Reed–Solomon code over $GF(512)$ concatenated with the (256, 9) binary Reed–Muller code.

7.8.2 *Decoding complexity*

We now take a rough and ready look at the complexity of decoding, that is the amount of computing effort needed by the decoding algorithm, and more specifically how it varies with n, the number of symbols in a codeword, as n becomes very large. (The complexity of encoding is usually much less than the complexity of decoding.) We first consider the outer decoder. We assume that the information rate k/n and the error rate t/n remain constant. Here k is the number of information symbols and t the number of errors. (For simplicity we assume that the inner decoders cannot declare erasures.) The number of coefficients in the syndrome polynomial $S(z)$ is $r = n - k$, which is of order n. Each coefficient involves a sum over codeword symbols, that is n calculations, and so overall to find $S(z)$ needs a number of calculations of order n^2. The Euclidean algorithm involves a long division on each iteration, but usually this is only a two-step process with a first-degree quotient emerging. Thus with a number of iterations of order n the number of computations is of order n^2. Finding the roots of the error locator polynomial by trial-and-error also involves a like number of computations, as does finding the error values y_k, so the overall complexity is of order n^2. This conclusion may be modified by the availability of 'fast' algorithms for finding S and the roots of Q, and for the Euclidean algorithm (Justesen 1976, Sarwate 1977). There is also the problem of the complexity of the basic Galois-field arithmetic, but this can usually be performed using whole field elements together, rather than on a bit-by-bit basis, just as ordinary integer arithmetic on a computer is not broken up into bit-by-bit computations, at least not on the programming level. Thus overall the complexity of the outer decoding process is of the order of a low power of n.

A typical inner-code decoding at worst would involve testing $2^m \sim n$ possible codewords against the received string. Each test would involve a sum along the codeword, thus requiring of the order of m computations. This has to be done n times over, once for each symbol of the Reed–Solomon code. The overall complexity is then of order $n^2 \log_2 n$, if we set $m = \log_2 n$. Thus we find that the total complexity varies as a low power of n, and not exponentially as in a minimum-distance decoding algorithm (Section 5.3.5). Thus values of n in the range of hundreds or thousands are feasible.

7.8.3 *Reliability*

We first consider the error probability of the outer Reed–Solomon decoder, as n becomes very large, with the information rate $R = k/n$ and the symbol-error probability p_{se} remaining fixed. (As before it is assumed that the inner decoders cannot declare erasures.) An error occurs if the number of symbol errors t exceeds $\frac{1}{2}(n-k)$, so that $t/n > \frac{1}{2}(1-R)$. Now if p_{se} is fixed at a value less than $\frac{1}{2}(1-R)$, then by the Chernoff bound (Section 4.1.4) it is exceedingly unlikely that t/n exceeds $\frac{1}{2}(1-R)$. The error probability falls off roughly exponentially with n.

Next we consider the inner decoders. Provided the channel is being operated at a fixed rate below its capacity, Shannon's device of averaging the error probability over all codes guarantees that there exists a reliable binary code with a maximum-likelihood decoding algorithm for which the word-error probability falls off as $2^{-\alpha m}$, where m is the number of bits to be encoded. The error exponent α is a positive quantity which becomes very small if the information-rate is close to the channel capacity. Since the inner code's word-error probability is the outer code's symbol-error probability we find p_{se} is of order $n^{-\alpha}$, so that it is not just constant as was previously assumed, but actually falls slowly with n increasing.

Thus we have the very important result that there are concatenated codes for which the decoding complexity is proportional to a low power of n, while the error-probability falls exponentially with n. In fact this can be achieved at any overall information rate below the channel capacity, this rate being the product of the inner-code rate and the Reed–Solomon rate R. The point is that since p_{se} falls off with n we can choose a value of R as close as we please to unit, say 0.99, and by choosing n larger enough we can arrange that p_{se} is less than the critical value 0.005. Thus if we wish to operate at 98.01 per cent of capacity we could operate the inner code at 99 per cent of its capacity, and the outer code at $R = 0.99$, provided we are permitted to make n sufficiently large. Thus Shannon's challenge is answered to a large extent.

Unlike the outer code the inner code is not completely specified. From a practical point of view good inner codes can be found either as members of well-understood families of codes, or by computer search through randomly chosen encodings. The inner codes need not be block codes at all, but convolutional or trellis codes for which the Viterbi algorithm provides a good algorithm for maximum-likelihood decoding. It seems unlikely that m can be much greater than 16.

From a mathematical point of view this is not a proper construction. Recently Delsarte and Piret (1982) found a very ingenious way of using Shannon's averaging procedure constructively. A different (p, m) code is chosen quite explicitly for each of the inner encodings of Fig. 7.2. These codes are chosen so that although it is not practicable to compute the error probabilities for each one, it is possible to use Shannon's methods to find the error probability averaged over all n of them. This error probability can be shown to go to zero as before as m becomes large. This is quite sufficient, since as far as the outer decoder is concerned the number of errors in an n-symbol codeword depends statistically only on this average. There is no need even to weed out the bad inner codes.

One outstanding problem remains. Shannon and Gallager have produced formulae giving bounds on the codeword lengths for a given reliability, and it is found that the codeword lengths of concatenated codes for a given reliability are much larger than what is theoretically necessary. There is a hint of this in Fig. 6.5 where the curve for code Z with its enormous codewords is still quite a long way to the right of the critical vertical at $\gamma_b = 0.693$. Thus there may be still better easily decodable codes.

Exercises

7.1 Write down the addition and multiplication tables for Z_7. Verify that every element other than 0 has an inverse. Which elements have square-roots? Find the primitive elements in Z_7.

7.2 Solve the linear simultaneous equations $2x + y = 4$, $2x - 5y = 1$ in $GF(7)$.

7.3 Solve the quadratic $x^2 + 6x + 5 = 0$ in $GF(7)$ (a) by trial and error, (b) by the formula for solving quadratics.

7.4 The field $GF(32)$ can be set up by putting into $GF(2)$ an element β satisfying $\beta^5 = \beta^2 + 1$. Set up the table of powers of β analogous to Table 7.3 and find the Zech logarithms. (Note for interest: Both α ($\alpha^4 = \alpha + 1$) and β can be added into $GF(2)$. The ensuing field is $GF(2^{20})$, a rather large one.)

7.5 Verify that (110 110 101 010 101) is a codeword in the (15, 11) Hamming code of Section 7.2.1, and correct the error in (101 010 010 000 111), assuming that there is only one.

7.6 Write down the binary check-sum conditions analogous to (7.4) implied by (7.6) and verify that they are merely linear combinations of the equations (7.4).

7.7 Let $s_j = \Sigma_i r_i \alpha^{ji}$ with r_i in $GF(2)$. Show that $s_2 = s_1^2$ and more generally that $s_{2l} = s_l^2$.

7.8 A codeword in the (15, 7) BCH code is received as (101 010 110 011 100). Use the algorithm of Section 7.2.2 to find the codeword, assuming that there are at most two errors.

7.9 How do the Hamming codes described in Section 6.3.4 and in Section 7.2.1 differ from each other?

7.10 A polynomial of degree m over a field F may be defined without any reference to an argument z as a string of $m + 1$ coefficients from F labelled from 0 to m. Write down the rules for adding and multiplying two polynomials

EXERCISES

defined in this way and show that multiplication is commutative and associative. (Hint: Worries over limits of summation may be avoided by introducing extra coefficients of zero value whose labels are out of range.)

7.11 Show that $z^4 + z + 1$ (over $GF(2)$) is irreducible, by considering all possible factors of degree 1 and 2. Why is it not necessary to consider factors of degree 3? Similarly show that $z^5 + z^2 + 1$ is irreducible, but that $z^5 + z + 1$ is not.

7.12 The derivative $A'(z)$ of a polynomial

$$A(z) = \sum_{k=0}^{n} a_k z^k$$

is defined as

$$A'(z) = \sum_{k=1}^{n} {}^{\iota}k{}^{\prime} a_k z^{k-1}$$

where 'k' is the sum of k 1s in $GF(q)$. Show that if $C(z) = A(z)B(z)$ then $C'(z) = A(z)B'(z) + A'(z)B(z)$. Also show that if $A(z)$ and $A'(z)$ are relatively prime then $A(z)$ does not contain a repeated factor.

7.13 Show that the remainder when $G(z)$ is divided by $(z - a)$ is $G(a)$. (This provides a practical way of evaluating polynomial expressions.) Hence show that $z - a$ is a factor of $G(z) - G(a)$ and that $G(a) = 0$ if and only if $z - a$ is a factor of $G(z)$.

7.14 Show that the number of roots of a polynomial cannot exceed its degree.

7.15 Show that if c is a root of any polynomial $A(z)$ over $GF(2)$, then so is c^2. *Hence* show that the roots of $z^4 + z + 1$ are α^2, α^4, and α^8, as well as α.

7.16 Show that if first-degree polynomials (of the form $az + b$) over the real field are added and multiplied mod $z^2 + 1$, then they act like complex numbers. What complex number corresponds to $az + b$? Use the Euclidean algorithm to find the inverse of $az + b$ mod $z^2 + 1$, with a and b not both zero.

7.17 Find the least value of n for which $J(z) = z^4 + z^3 + z^2 + z + 1$ is a factor of $z^n - 1$ (over $GF(2)$). Do the same for $J(z) = z^4 + z^2 + 1$.

7.18 Verify that $z^4 + z + 1$ is a factor of $z^{15} - 1$ (over $GF(2)$), and *hence* show that $\alpha^{15} = 1$.

7.19 Use Euclid's algorithm to show that $A(z) = z^4 + z + 1$ and $B(z) = z^4 + z^3 + z^2 + z + 1$ are relatively prime.

7.20 Assuming that there are at most two errors, decode the received word (111 010 111 000 000) for the (15, 7) code of Section 7.2.2, using the Euclidean algorithm. (Note that all the syndromes s_1, s_2, s_3, and s_4 are needed for $S(z)$.)

8. Optical communications and quantum effects

8.1 Optical communications

8.1.1 *Introductory*

Visible light is a form of electromagnetic radiation occupying the band of frequencies from about 400 THz to 700 THz (1 THz = 1000 GHz), five orders of magnitude larger than microwave frequencies. Interest in communications at such frequencies has grown enormously because of

(a) the development of coherent laser sources, completely analogous to coherent radio transmitters and

(b) the development of dielectric fibre waveguides (optical fibres) with very low losses at these frequencies.

The main advantages of using such high frequencies are

(a) available bandwidths are huge, even when the bandwidth is narrow compared with the carrier-frequency, and

(b) short wavelengths imply that directional antennae need not be very large in free-space communications, and also that in guided communications the waveguides are very thin fibres; thus many fibres can be included in one cable.

Terrestrial communication by electromagnetic waves at frequencies from roughly 300 GHz to 30 THz is ruled out by the fact that such radiation is capable of exciting molecular vibrations and rotations in almost any dielectric substance and hence is strongly absorbed. In consequence it is not possible to transmit such radiation either along waveguides or through the atmosphere. Above about 30 THz the frequencies are too high to excite molecular vibrations and so there is a band of frequencies which need not be strongly absorbed. Hence transmission along dielectric waveguides becomes possible. Transmission through the atmosphere is also possible but the radiation is very readily scattered by dust, clouds, rain, and so on, so that its use in the atmosphere is unreliable. Above 1000 THz in the ultraviolet radiation produces electronic transitions in most substances and so is again strongly absorbed. Signalling at frequencies between 30 THz and 1000 THz is referred to as optical communication. The usual method of transmission for terrestrial communications is along optical fibres, which are dielectric waveguides with a very narrow cross-section. Optical frequencies have also been proposed for data-links with deep-space probes (Pierce and Posner 1980).

In this chapter we pursue the theme of reliable digital communications, at optical frequencies. At these frequencies the quantum-mechanical nature of electromagnetic radiation manifests itself strongly, principally in two ways:

(a) Electromagnetic radiation can act as though it were a stream of particles (quanta, photons) and
(b) thermal noise is reduced so drastically that it can usually be neglected.

The former effect limits the amount of information that a beam of electromagnetic radiation can carry even if there is no extraneous noise (Garrett 1983).

This chapter is organized as follows: In the rest of Section 8.1 optical fibres are discussed as a transmission medium, to see what they can do and what may be hoped for. In Section 8.2 we discuss the problems of reception which are the principal causes of the limitations on information transmission. (Transmitters also have their problems, but these are practical, rather than fundamental.) Non-optimized systems are discussed first, and then we see how much further in principle their performance may be improved. We start by considering the basic ideas of photon detection (Section 8.2.1) and introduce simple unoptimized systems (Section 8.2.2), ideal detectors (Section 8.2.3), and finally coded systems (Section 8.2.4). These systems all use incoherent detection. In contrast coherent systems are discussed in Section 8.2.5, where a local oscillator is used at the detector.

Just as fundamental limits for information transmission over an AWGN channel are set by Shannon's formula, so there are fundamental limits for information transmission over a quantum channel. These limits are explored in Section 8.3 and we shall see the systems of Section 8.2 under ideal circumstances are not too bad in comparison.

8.1.2 *Optical fibres*

More than anything else, even the development of the laser, the production of cheap low-loss optical fibres has revolutionized optical communications to such an extent that most wide-band point-to-point wired communications will be taken over by such systems. The old coaxial and microwave wave-guide links are effectively superseded, and it is possible that even links for telephones and visual-display units within a single building may be replaced by optical fibres, not least because of their immunity from electrical interference and from problems caused by 'earth loops'.

A typical so-called 'monomode' optical fibre operates at a (vacuum) wavelength of 0.85 μm (i.e. at a frequency of 353 THz). It consists of a core of silica with a diameter of a few microns, surrounded by a cladding also of silica, but with a slightly lower refractive index. This cladding has a thickness of at least ten times the core radius. The radiation is kept close to the core by a phenomenon akin to total internal reflection.

The attenuation can be made as low as 1 dB/km by carefully purifying the silica. In practice this means that repeaters may be spaced by up to 30 km. The refractive index varies slightly with frequency, so that dispersion affects wide-band signals, and in consequence it is found that the usable bandwidth varies inversely with the distance of propagation. A typical fibre has a bandwidth-distance product of 40 GHz km. This can be improved by the use of optical equalizers, but the cost of doing this may well exceed the cost of using several fibres in parallel in the same cable, each operating with a reduced bandwidth, or the cost of using a shorter spacing between repeaters.

More generally it appears that optical fibres have so improved transmission capabilities that it is usually unnecessary to operate them anywhere near the theoretical limits. Thus the sources are often not coherent lasers, but incoherent light-emitting diodes. (Such devices might well be regarded as the optical-frequency analogues of spark-gap transmitters.) Such spatially diffuse sources do not couple efficiently into monomode fibres which, as the name suggests, are designed to propagate only one mode of the electromagnetic field, like wave-guides at microwave frequencies. Monomode fibres have to be driven by lasers. For incoherent sources *multimode* fibres are used, designed to operate with several wave-guide modes propagating together. Naturally the dispersion is much worse than for monomode fibres.

The choice of operating frequency in the vicinity of 0.85 μm (353 THz) was dictated by the availability of sources and detectors, and it is not optimal from the point of view of the fibre. For long-haul transmission the losses must be reduced as far as possible. One of the main causes of loss is the fundamental Rayleigh scattering, produced by the disorder in the medium associated with thermal molecular vibrations and compositional disorder on a microscopic scale inevitable in a glass. This scattering varies as the fourth power of the frequency. (Rayleigh scattering of sunlight in the atmosphere produces the blue colour of the sky and the red colour of sunsets.) Hence it is advantageous to decrease the frequency. For silica fibres an optimum occurs at around 1.3 μm (231 THz). Below this frequency the losses start to rise again because of the effect of molecular vibrations. The minimum attenuation available is around 0.5 dB/km, and is also fortunately associated with a minimum in the dispersion.

To attain still lower losses it has been proposed that fibres made out of heavy-atom compounds such as thallium bromide be used. The greater masses of the atoms compared with those of silicon and oxygen in silica cause the natural molecular vibrational frequencies to be much lower. Hence lower frequencies may be used, with lower Rayleigh scattering losses. Theoretically losses of below 10^{-2} dB/km are possible at around 4 μm (75 THz). Such low values would make transatlantic cables feasible without repeaters. (For further information on optic fibres see Howes and Morgan 1980, Keiser 1983, Midwinter 1979, Mimms 1982, or Personick 1981.)

8.2 Reception

8.2.1 *The photo-electric effect*

A beam of monochromatic electromagnetic radiation of frequency f can act as though it were a stream of particles (quanta, photons) each with a fixed amount of energy which is given up when the particle is absorbed and destroyed. This energy is hf, where f is the frequency of the radiation and h is Planck's constant, 6.626×10^{-34} J/s. At microwave and millimetre frequencies these quanta have an exceedingly tiny energy but at optical frequencies the energy is large enough for the granularity to show in some circumstances. Quantum limitations on information rates dominate noise-limitations at carrier-frequencies f satisfying $hf > k_B \theta_N$ where θ_N is the noise temperature and k_B is Boltzmann's constant. According to this criterion we should not worry about quantum effects at frequencies much below 6THz if the noise temperature is 300 K.

One manifestation of the quantum nature of light is the photoelectric effect. If a beam of light falls on the suitably prepared surface of a metal or semiconductor in vacuum, then the energy hf of an absorbed photon will be sufficient to eject an electron from the metal surface into the vacuum provided that hf exceeds the binding energy W which normally keeps the electron in the metal or semiconductor. These ejected electrons may be picked up by a positive electrode (anode) in the vicinity and so an electric current flows. Such an arrangement is called a vacuum photodiode. This *photoelectric effect* evidently only occurs above a critical frequency W/h regardless of the intensity of the incident light. Moreover the number of photons arriving each second is given by the power P divided by hf, so that the maximum electric current carried by these electrons into the vacuum is equal to $qP/(hf)$, where $q = 1.602 \times 10^{-19}$ coulombs is the magnitude of the charge on an electron. As we might expect not all the absorbed photons eject electrons and we therefore must multiply the current by a 'quantum efficiency factor' η_Q. In practice η_Q can have quite substantial values, around 0.3 to 0.6.

Semiconductor photodiodes operate in an analogous manner. A slab of intrinsic semiconductor is sandwiched between two electrodes which maintain an electric field in it. A photon is absorbed in the slab, producing an electron-hole pair. The electron and hole are then swept out (in opposite directions) by the electric field, and their motion produces a pulse of current in the circuit maintaining the field, just as in the vacuum photodiode.

The total charge-transfer (in both cases) is the electron charge q and gives rather a weak pulse. Simple amplification of the current pulse is possible, but it introduces noise which may be excessive. An alternative is to multiply the number of carriers produced by each photon. This is carried out as follows in the semiconductor photodiode. The applied electric field is made so strong that the liberated electron (or hole) acquires so much energy before reaching the electrode which is its goal that it becomes capable of producing further electron–hole pairs. These newly liberated particles in their turn create further pairs, and so on. The overall effect is to produce a substantial pulse of current. A diode operated in this manner is called an 'avalanche' photodiode. In a vacuum photomultiplier the liberated electron is made to strike an anode from which it knocks out several more electrons, which are then accelerated until they strike another anode at a voltage higher than the first anode. This process is repeated several times.

Finally it should be mentioned that the larger the photon energy the more easily is this kind of photodetection carried out. The design of photodiodes operating at frequencies at and below 1.3 μm (231 THz) is not altogether a simple business.

8.2.2 Photodetection followed by amplification

We now consider briefly the kind of signal level needed at the detector to achieve reliable communication. It will be assumed that binary on-off keying (OOK) is used. The maximum tolerable bit-error rate is usually taken to be 10^{-9}. The first-stage amplifier is typically a field-effect transistor with a certain input capacitance C. We suppose that this capacitor is charged up through a resistor and is partially discharged through the photodiode each time a pulse of light arrives. The amplifier produces noise represented by a temperature θ_N. The uncertainty in the voltage as measured by the amplifier is equivalent to a charge $\bar{Q}$ given by equating the energy of the capacitor with this charge to half the noise energy:

$$\tfrac{1}{2}\bar{Q}^2/C = \tfrac{1}{2}k_B \theta_N$$

(Reif 1965 Section 6.6). Here k_B (1.38×10^{-23} J/K) is Boltzmann's constant. Thus we find

$$\bar{Q} = \sqrt{(Ck_B\theta_N)}.$$

The total charge transferred by the pulse must greatly exceed this quantity and hence we find that the number N of photons needed (producing a charge-transfer of Nq) must satisfy

$$N \gg q^{-1}\sqrt{(Ck_B\theta_N)}.$$

For typical values of $C = 1$ pf and $\theta_N = 1000$ K we find that this gives $N \gg 750$. In practice for reliable communications we need of the order of 10^4 photons for each bit which, with a photon energy of 1.53×10^{-19} J (at 1.3 μm) implies 15 μW at 1G bit/s (-18.2 dBm). A typical solid-state laser produces 100 μW (i.e. -10 dBm) so that an attenuation of 8 dB is permissible, giving a propagation distance of 16 km if the fibre-attenuation is 0.5 dB/km.

We can improve the range by using a more powerful source. The transmitter power is limited to the order of 1 W by the need to avoid non-linear effects in the fibre. Gas lasers of this power are available but they are bulky, fragile and expensive. (Of course the coherence of the output radiation means that almost all the output may be directed into the thin fibre in spite of the large size of the source.)

The use of avalanche photodiodes can improve the sensitivity significantly. Silicon avalanche photodiodes at 0.85 μm have been reported to give a bit-error rate of 10^{-9} with a few hundred photons per bit (Howes and Morgan 1980, p. 14). Unfortunately these efficient devices are not effective at 1.3 μm, and the increased sensitivity is vitiated by the higher loss in the fibre caused by having to operate at the higher frequency. Improved materials and techniques are rapidly being developed for 1.3 μm and lower frequencies.

8.2.3 *The ideal photodetector*

An ideal photodetector should produce a strong and very brief pulse of current every time a photon strikes it. Real photodetectors are deficient in at least two respects:

(a) Not every photon causes a response, but only a fraction η_Q of those arriving, where η_Q is the quantum efficiency. We may simulate this deficiency by preceding our ideal photodetector by an attenuator, which only allows a fraction η_Q of the power through.

(b) A detector can produce a pulse even if a photon does not arrive. Such an event may be called a 'false alarm'. Thus an electron may be ejected from the cathode of a vacuum photodiode by thermal vibrations. If the energy W required to eject an electron is very much greater than the thermal energy $k_B \theta$, then the ejection rate varies roughly as $\exp(-W/k_B \theta)$ and may be heavily reduced by cooling the photocathode.

If steady monochromatic radiation at frequency f and with power P falls on an ideal detector, then as described in Section 8.2.1 the detector responds as though there is a stream of photons (each of energy hf) arriving independently at a mean rate

$$\lambda = P/(hf). \tag{8.1}$$

The situation is analogous to that described in Section 4.2.5 in connection with shot-noise.

Suppose that a pulse of such radiation of duration τ falls on the detector. We divide up this interval into a huge number M of equal subintervals each of duration τ/M, and suppose that the probability of a photon arriving in a given subinterval is $\lambda \tau/M$ (provided that M is so large that this quantity is very small). At the end we let M tend to infinity. It is quite possible that no photons at all are detected during the arrival of the pulse. The probability of finding no photons in a given subinterval is $(1 - \lambda\tau/M)$ and because of the independence of the arrival times of the photons the probability p_0 of finding no photons at all is the product of M factors each with this value. Thus we find that

$$p_0 = \lim_{M \to \infty} (1 - \lambda\tau/M)^M = \exp(-\lambda\tau). \tag{8.2}$$

The quantity $\bar{N} = \lambda\tau$ is (as it should be) the mean number of photons arriving in the pulse, so we may write

$$p_0 = \exp(-\bar{N}). \tag{8.3}$$

More generally it can be shown that the probability p_n of finding n photons in the pulse is given by

$$p_n = \bar{N}^n \exp(-\bar{N})/n!. \tag{8.4}$$

Let us now consider a simple signalling system. Photon detectors have no way of telling the phase of the received radiation, for reasons discussed below. Hence systems of modulation such as BPSK and QPSK cannot be used with direct photon-detection, and the simplest method of all, that of on-off keying (OOK), is the obvious choice. During the off-pulses nothing is transmitted, and provided there are no 'false alarms' the detector will not respond. Hence the off-pulses are faithfully recorded. This is not so for the on-pulses, since there is a probability given by (8.3) that nothing will be detected. This may be interpreted variously as follows: in free-space communications all the photons sent out in a huge burst by the transmitter have spread out so thinly by the time they reach the receiver that they all miss the detector completely; in guided communications through an absorbing medium they all happen to have been absorbed before reaching the detector. Using (8.3) we find that to make p_0 less than 10^{-9} the value of $\bar{N}$ must not be less than 20.72. Thus if on- and off-pulses occur with equal frequency the number of photons per bit of information needed to achieve this level of reliability is just under 10.5. This is about 30 dB better than the system described in the previous section.

8.2.4 *PPM system*

To do better we must use some form of coding. A theoretically good system was suggested by Pierce (Pierce and Posner 1980), and is analogous to the PPM systems of Section 4.3.4. It is an effective system when the signals are power-limited rather than band-limited, which is what may be expected in free-space communications. Biorthogonal signalling was mostly discussed in Section 5.2.4, since the sign of the pulse could be used to convey an extra bit of information. In the present context this is ruled out since the phase of the received carrier is completely indeterminate.

Let us assume that we can send B pulses per second, so that time is divided into slots of duration B^{-1}. The mean received power is P, so that the mean number of photons per slot $\bar{n}$ is given by

$$\bar{n} = P/(hfB). \tag{8.5}$$

Another interpretation of $\bar{n}$ is the photon-rate divided by the pulse-rate B. (It will be seen in Section 8.3.4 that B is indeed the quantum analogue of the bandwidth equivalent of Section 5.1.2.) In a power-limited situation B may be large, so that $\bar{n}$ can be very much less than 1. Under these circumstances we proceed as follows: the slots are grouped into sets of Q adjacent slots, and the transmitter is quiescent in all the slots except one, when a large burst is fired. In this way we signal our choice of one symbol out of an alphabet of size Q. Since the photons are all saved up for the one favoured slot, the mean number of photons received in that slot is $Q\bar{n}$ where $\bar{n}$ is the overall mean number of photons per slot. The receiver simply detects the arrival of photons without counting them. Therefore it reliably finds the unused slots, but there is a probability $\exp(-Q\bar{n})$ that all the photons in the pulsed slot miss the detector. This result follows from (8.3) with $\bar{N}$ replaced by $Q\bar{n}$. Hence the receiver does not know what was sent in that group and has to declare an erasure. The scaled information rate $J' = J \ln 2$ is given by

$$J' = (B/Q) \ln Q \{1 - \exp(-Q\bar{n})\}, \tag{8.6}$$

and the factors in this expression arise as follows. The first factor (B/Q) is the rate at which groups arrive. The second factor $\ln Q$ is the scaled amount of information conveyed by each group, and the third factor is the reduction in the rate caused by the erasures. It is perhaps surprising that the effect of the erasures on the information-rate is just a reduction by the fraction of erasures. This is a result following from Shannon's formula for the capacity of a so-called 'discrete memoryless channel' (Rosie 1973). A suitable coding scheme which enables us to approach this information rate as closely as we please is discussed below.

We may optimize this rate with respect to Q, but a reasonable choice is to set $Q = 1/\bar{n}$, which gives a scaled information-rate of

$$J' = 0.673 \, B\bar{n} \ln(\bar{n}^{-1}).$$

(Since photons arrive at a mean rate $B\bar{n}$ the information per photon is of the order of $\log_2(\bar{n}^{-1})$ bits, which can theoretically be made as large as we please, but in practice could not exceed, say, 10. Thus there is a limit of 0.1 photons per bit, considerably smaller than the value of 10^4 for the system described in Section 8.2.2. See Lesh 1982.)

To obtain reliable communications in the presence of erasures we may use codes like the Reed-Solomon codes of Section 7.5. Such a code operates over a field of size Q (so that Q must be a prime power). An (N, K) code (over $GF(Q)$) can fill in $(N - K)$ erasures. Thus we use a form of concatenated code.

The input block to the Reed–Solomon encoder consists of K symbols from an alphabet of size Q. The output is N symbols also from this alphabet. Each symbol is then encoded again by the modulator or inner encoder as a single pulsed slot chosen out of a block of Q slots. (Thus altogether QN slots are used to send the complete block.) We note that the rate is reduced by a factor

$$R = K/N$$

compared with the rate without the Reed–Solomon encoder.

Let p_0 be the probability of erasure. Then the probability that in N symbols there are more than $N(p_0 + \epsilon)$ erasures falls exponentially with N, for any fixed positive quantity ϵ (Section 4.1.4). Thus if

$$N - RN > N(p_0 + \epsilon),$$

that is if

$$1 - R > p_0 + \epsilon,$$

the error probability falls exponentially with N. So provided R is less than $1 - p_0$ we can fill in the erasures with a reliability as high as we please by taking N large enough.

We should note two differences between this system and the PPM system described in Section 4.3.4 for dealing with Gaussian noise:

(a) In that system we obtain optimal results by letting Q tend to infinity, which is fortunately not the case here.

(b) In that system we could use phase information for increasing the rate marginally. In the photon counting system this is not possible.

8.2.5 *Coherent reception*

The phase of the carrier within a pulse may be measured if the incoming signal at frequency f_C is mixed with the signal from a local oscillator at a nearby frequency f_L and the two beams are allowed to fall on to a photodiode. The curious nature of quantum theory enables us to calculate the response as follows (provided that f_C/f_L is very close to unity): the power $P(t)$ of the radiation falling on the diode is calculated from classical wave-theory and hence shows beats at the difference frequency $f_I = (f_C - f_L)$. The probabilistic rate of arrival of photons is then given by $P(t)$ divided by the photon-energy. Moreover the photons arrive independently. We shall take the photon-energy as hf_L, ignoring the difference between the energies of the signal and local-oscillator photons.

A schematic diagram of the receiving system is shown in Fig. 8.1. The signal is first passed through a narrow-band filter to exclude radiation at frequencies outside the band of interest. Then it is mixed with the radiation from a local

RECEPTION

[FIG. 8.1 Optical heterodyne receiver — diagram showing narrow-band filter, partial reflector, local oscillator (lo), photon detector, amp with gain $\times g$, multipliers with $\cos 2\pi f_1 t$ and $\sin 2\pi f_1 t$, and integrators $\int_0^{T_0}$.]

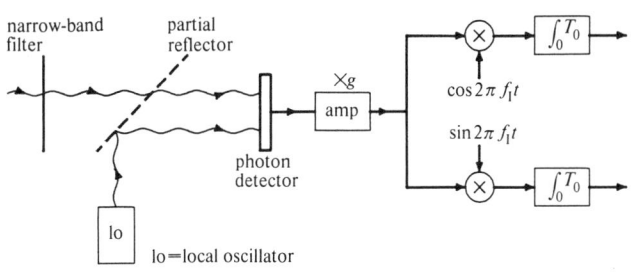

FIG. 8.1 Optical heterodyne receiver.

oscillator using an almost transparent partially reflecting mirror, and then the two beams are allowed to fall on to a photodiode. (It is important that the two beams are parallel at the detector, for otherwise beats from one part of the photon-counting surface are out of step with those from another part, and so cancel out.) Then the electrical output is amplified, split, and sent to in-phase and quadrature detectors, after which on each channel it is integrated over the duration of a pulse. During the duration of a given pulse from $t = 0$ to T_0 the signal is taken to be

$$s\sqrt{(2/T_0)}\cos(2\pi f_C t - \theta)$$

with magnitude s and phase θ, and the local oscillator signal is

$$s_L\sqrt{(2/T_0)}\cos(2\pi f_L t)$$

with $s_L \gg s$, as is usually the case. These expressions are scaled so that s^2 and s_L^2 are the total energies over the duration of the pulse. The beat frequency $f_1 = f_C - f_L$ is very much less than f_C or f_L. It is assumed that f_C exceeds f_L slightly. The power is the square of the sum. The detector is not fast enough to follow oscillations at optical frequencies or higher, but it can follow the beats at the beat frequency, which may well be in the microwave range. Hence the probabilistic rate of detection of photons is equal to

$$\lambda(t) = \{\eta_Q/(hf_L)\}\{s_L^2 + 2ss_L\cos(2\pi f_1 t - \theta)\}/T_0, \qquad (8.7)$$

with η_Q the quantum efficiency. Here we have neglected a term containing s^2, since $s_L \gg s$. Whenever a photon is detected at the instant t it causes a pulse of current to flow which if it were not for the amplifier and multiplier would increment the charge on the integrator by q, the charge on an electron. However this is changed to

$$C(t) = gq\cos(2\pi f_1 t) \qquad (8.8)$$

on the in-phase channel, where g is the amplifier gain, and similarly with sin replacing cos on the quadrature channel.

We break up the total integrating time T_0 into a huge number of infinitesimal time-slots of duration Δ. The probability of a 'hit' in the kth slot is $\lambda(k\Delta)\Delta$, and it follows from quantum theory that 'hits' and 'misses' in different infinitesimal slots are independent of one another. We now define the random variable S_k as the 'score' from the kth slot given by

$$S_k = \begin{cases} C(k\Delta) & \text{if 'hit'} \\ 0 & \text{if 'miss'} \end{cases}$$

The total integrated output is the sum of these scores

$$S = \Sigma_k S_k.$$

Moreover the scores in different slots are independent of one another so that the mean and variance of the total are given by the sums of the means and variances respectively of the S_k. The mean of S_k is simply $C(k\Delta)\lambda(k\Delta)\Delta$, and so the mean of the total score S is

$$\langle S \rangle = \Sigma_k C(k\Delta)\lambda(k\Delta)\Delta \to \int_0^{T_0} C(t)\lambda(t)\,dt,$$

on replacing the sum over the infinitesimal slots by an integral. We assume that the interval T_0 is a whole number of periods at the intermediate frequency f_I, so this integral is by (8.7) and (8.8)

$$\langle S \rangle = \{s_L g q \eta_Q/(hf_L)\} s \cos\theta. \tag{8.9}$$

Thus the mean of this output is proportional to the in-phase component $s\cos\theta$ of the signal. We may similarly obtain the quadrature component.

The noise is the variance of S which is given by

$$\text{var}(S) = \Sigma_k \text{var}(S_k),$$

with

$$\text{var}(S_k) = \langle S_k^2 \rangle - \langle S_k \rangle^2.$$

But we see that

$$\langle S_k^2 \rangle = \{C(k\Delta)\}^2 \lambda(k\Delta)\Delta,$$

while $\langle S_k \rangle^2$ is negligible, being of second order in Δ. Thus after replacing the sum over k by an integral we find that

$$\text{var}(S) = \int_0^{T_0} \{C(t)\}^2 \lambda(t)\,dt.$$

The only contribution to this integral comes from the first term in $\lambda(t)$ (8.7) and so gives

$$\text{var}(S) = \tfrac{1}{2}\{\eta_Q/(hf_L)\} g^2 q^2 s_L^2. \tag{8.10}$$

If s_L is so large that many photons are detected during the pulse, then S is the sum of a large number of independent variables and its distribution is approximately Gaussian.

The noise-energy summed over both channels is twice this. The signal-energy in the first output is the square $\langle S \rangle^2$ of the mean signal as given by (8.9), and so the total signal energy is

$$\{s_L g q \eta_Q/(hf_L)\}^2 s^2 (\cos^2\theta + \sin^2\theta).$$

Thus the overall signal-to-noise ratio is $\{\eta_Q/(hf_L)\}s^2$. Since f_L and f_C are almost the same this to a good approximation is $\eta_Q \bar{n}$ where $\bar{n}$ is the number of signal-photons in the pulse. So by Shannon's formula (Section 5.3.2) the capacity is $B \log_2(1 + \eta_Q \bar{n})$ where B is the pulse-rate.

The coherent system thus acts as an AWGN channel with an 'η-value' or noise-energy given by

$$\eta_{NE} = hf/\eta_Q.$$

(Here η_{NE} is the η of previous chapters.) In other words the noise-temperature θ_N is given by

$$k_B \theta_N = hf_C,$$

that is, 10 000 K at 200 THz if $\eta_Q = 1$. Coherent detection is of interest because:

(a) The sensitivity is reasonably high. The noise-temperature sounds appalling, but it corresponds to 1 photon per pulse, so that high reliability could be attained for, say, 20 photons per bit.

(b) The detection process provides amplification, so that noise from subsequent amplification is not so serious. This is because the output (8.9) is proportional to the amplitude of the local oscillator.

(c) If necessary equalization to compensate for optical dispersion can be carried out electronically at the beat frequency.

Laser preamplifiers may also be used in conjunction with coherent detection, but the noise performance of the ideal system is no better than the noise-performance of the ideal coherent system without preamplification. However, preamplifiers are useful if the quantum efficiency of the detector is low. Since lasers suffer from spontaneous emission they should not normally be used in conjunction with incoherent detection.

8.3 Quantum theory

8.3.1 *Photons in an ideal single-mode cavity*

It is now necessary to examine some of the basic ideas in quantum theory to see why there is a limit on the rate of transmission of information. Fortunately these ideas can be presented without the need for a detailed knowledge of quantum theory, although in consequence the treatment may be non-rigorous and a bit sketchy in parts. (For an introduction to quantum theory see Wichmann 1972.)

We start by considering the oscillations of the electromagnetic field in an ideal tuned cavity which supports only one mode of oscillation of interest at a frequency f. The state of the electromagnetic field may then in classical theory be specified by giving the amplitude and phase of this oscillation, or equivalently the in-phase amplitude A' and quadrature amplitude A'' as measured relative to some master-oscillator or clock at the same frequency. (It will be assumed that these amplitudes are scaled so that the sum of their squares is the total energy.) These variables may take on a continuum of values, subject only perhaps to a condition that the total energy should not exceed some limit E_{max}. The fact that the radiation in the cavity is quantized shows up when we attempt to determine the energy of the oscillating field. It is found that the energy is always an integral multiple of hf. We also find limitations on the measurements that can be made. In particular if having measured the energy of the oscillation we measure the phase it is found that the answer is totally unpredictable (because the first measurement has changed the quantum state of the system to a state where the phase is undefined). This is a manifestation of the Heisenberg 'uncertainty principle' which in this case states roughly that the precision Δn with which we measure the number of photons and the precision $\Delta\phi$ with which we may measure the phase ϕ are limited by $\Delta n \, \Delta\phi \approx 1$. So a precise knowledge of the number of photons ($\Delta n = 0$) completely precludes any knowledge of the phase. Conversely if we have set up the oscillation in a state where there is some knowledge of the phase, then a subsequent measurement of the number of photons will always show some scatter, although it will always give a non-negative integer. The in-phase and quadrature amplitudes A' and A'' are also subject to a similar relationship which states that the uncertainties $\Delta A'$ and $\Delta A''$ respectively in their measurements must satisfy $\Delta A' \, \Delta A'' \approx hf$. In particular if these uncertainties are equal then they satisfy

$$\Delta A' = \Delta A'' \approx \sqrt{(hf)}. \tag{8.11}$$

The theory does not specify how these measurements are to be made, but it is reasonable to expect that they are idealizations of the two types of detection described so far, incoherent 'photon-counting' detection (Section 8.2.3), and 'coherent' detection (Section 8.2.5). Thus to count the photons in the cavity we may couple the mode to a photodetector, the number of pulses from which determines the number of photons. This method of making the measurement changes the state of the cavity to a state with no photons at all.

8.3.2 The suppression of thermal noise

The field in the cavity may be excited by thermal radiation from a source at temperature θ_N. This excitation is just thermal noise. At low frequencies ($hf \ll k_B\theta_N$) the energy has an average value $k_B\theta_N$, but at high frequencies the chaotic thermal excitations find it hard to get together sufficiently to put together the energy needed to produce a photon of energy greater than $k_B\theta_N$. Hence the noise-energy is reduced. According to a general principle of statistical mechanics the probability of finding the cavity with n photons in it falls exponentially with the energy nhf and is given by $p_n = Z^{-1}\exp\{-nhf/(k_B\theta_N)\}$ (Reif 1965, Section 4.5). Here Z is the normalization factor which determines that the probabilities sum up to unity, and so is given by summing a geometric progression:

$$Z = \Sigma_{n=0}^{\infty}\exp\{-nhf/(k_B\theta_N)\} = [1 - \exp\{-hf/(k_B\theta_N)\}]^{-1}. \quad (8.12)$$

The mean energy is thus $\Sigma_n(nhf)p_n$. By differentiating (8.12) with respect to f this expression can be evaluated as

$$\bar{E} = hf/[\exp\{hf/(k_B\theta_N)\} - 1]. \quad (8.13)$$

It may be readily checked that at low frequencies when $hf \ll k_B\theta_N$, the value of $\bar{E}$ is approximately equal to $k_B\theta_N$, but that as f tends to infinity $\bar{E}$ falls rapidly to zero. The mean photon-number is $\bar{E}/(hf)$. At optical frequencies this value is very small indeed for $\theta_N \leq 300$ K, which is why thermal noise has not been considered in this chapter. (One practical precaution may be needed. Any detection system should be preceded by a narrow-band optical filter to remove photons at frequencies different from those used for signalling. This filter may have to be cooled, for otherwise it may radiate more photons than it absorbs!)

8.3.3 *Maximum rate of transmission of information*

Suppose now we wish to use the state of the field in the cavity as a way of conveying information. We may picture the transmitter as setting up the field with specified values of A' and A'' which are chosen according to some code. The cavity is then sent to the receiver which measures these amplitudes and decodes the signal. According to the classical or pre-quantum theory the amount of information which can be sent is in principle infinite since the amplitudes may be specified with arbitrary precision. In the quantum scheme this arbitrary precision is no longer available because of the uncertainty principle. One method of encoding the cavity-field is to use the number of photons n to represent the message. Any attempt to send more information by setting the phase as well fails, because this process alters the photon number by an indefinite amount. So choosing a value of n is the best that can be done and with an upper bound on the energy of E_{max} the number of values of n cannot exceed $E_{max}/(hf)$. (For the sake of the argument it is assumed that this is a fairly large number.) So the information that can be sent in this way is just $\log_2\{E_{max}/(hf)\}$ bits. This is a scheme using photon-counting measurements. An alternative encoding scheme based on 'coherent' measurements is one rather like the signal constellations previously described, where the transmitter spaces the signal points

(A', A'') in the complex plane with a spacing of at least $\sqrt{(hf)}$. Then by (8.11) the receiver can determine with reasonable reliability which signal-point was actually used. The restriction on the energy requires that all these points lie within a circle of radius $\sqrt{(E_{max})}$, and so the total number of points is very roughly the ratio of the areas of a circle of radius $\sqrt{(E_{max})}$ to that of a circle of radius $\sqrt{(hf)}$, that is $(\pi E_{max})/(\pi hf)$. Taking the logarithm of this quantity gives an estimate for the information of $\log_2 \{E_{max}/(hf)\}$ as before. Thus we find that the information is restricted even without having to invoke any form of noise.

So far only a single mode has been considered. Let us now suppose that the signal at the receiving end has been fed into a wave-guide loop (Section 2.3.5). The wave-guide is just long enough to accept the whole signal, whose duration is T. As soon as the signal is inside, the input of the guide is quickly switched from the receiving antenna to the output of the guide, which is looped round so that the output is close to the input. The signal now circulates indefinitely with period T. We now investigate the information content of the field in this looped guide. The loop is also a form of cavity, but with many modes of vibration. The ones of interest are the waves travelling in the right direction, having the correct polarization, and lying in the right frequency-band. Each mode has to have an integral number k of wavelengths round the loop. If we use this number k to label the modes it follows that the kth mode has the frequency $f = k/T$. In the classical theorem the amplitude of each mode may be specified by two real numbers A'_k and A''_k, or by the complex number $A_k = A'_k + jA''_k$, and the amplitude of the field at some point in the loop may then be written as a Fourier series as in Section 2.3:

$$s(t) = \Sigma_{k(>0)} A_k (2T)^{-1/2} \exp(2\pi jkt/T) + \text{complex-conjugate}. \quad (8.14)$$

How could this loop be used for conveying information? Let us assume that the average power being sent is P, so that the total energy is PT. Let us also suppose that the signals occupy a bandwidth B about a carrier frequency f so that only BT modes are used. Moreover we shall assume that the bandwidth B used is very much less than f_C so that the energy of any photon may be taken as having the same value hf_C. Thus we have a number $M = PT/(hf_C)$ photons to be distributed among $N = BT$ modes. Each distinct distribution corresponds to a possible signal, and it is an easy combinatorial problem to show that there are $(M + N - 1)!/\{M!(N - 1)!\}$ distributions. The information I is then given by the logarithm of this quantity, and for this we use Stirling's formula which to an accuracy sufficient for our purposes states that if n is large then $\ln(n!) \approx n \ln(n) - n$ (Exercise 3.4). Hence the scaled information $I' = I \ln 2$ is given by

$$I' = BT\{(\bar{n} + 1) \ln(\bar{n} + 1) - \bar{n} \ln(\bar{n})\}$$

where

QUANTUM THEORY

$$\bar{n} = P/(Bhf) \tag{8.15}$$

is the average number of photons in each mode of the loop at the receiver. So dividing by T we have a scaled information rate

$$J' = J \ln 2 = B\{(\bar{n} + 1) \ln (\bar{n} + 1) - \bar{n} \ln (\bar{n})\}. \tag{8.16}$$

It must be pointed out that this is an upper bound on the information rate which cannot be achieved even in theory on free-space links, ultimately because the remote transmitter does not have complete control over what is received. Moreover it is derived subject to the assumption that $B \ll f_C$ and so the conclusion that J tends to infinity when B tends to infinity keeping P fixed is invalid. It is possible to derive answers for the wide-band case by using counting techniques borrowed from statistical mechanics. In the extreme case when there is no restriction at all on the bandwidth it can be shown that an upper bound on the information rate is

$$C_{max} = 3.7007 \sqrt{(P/h)}. \tag{8.17}$$

This formula may be made plausible as follows. To send information at a rate C we need about C pulses per second. This requires a band of frequencies from $f = 0$ to $f = C$ roughly. So a typical photon-energy is of order hC. Each pulse must contain at least one photon to be detectable so the total power P is $C(hC)$. Solving $P = hC^2$ for C gives the above result, except for the numerical factor.

8.3.4 More on photon counting

An idealized method to determine the numbers of photons in each mode in the loop may be pictured as follows. The radiation in the loop is fed out as a parallel beam by an antenna in the reverse way to the way it was set up, and then fanned out into its different frequency components by a diffraction grating or prism, just as white light can be broken up into its separate colours. Each frequency component then goes to an ideal photodetector which counts the number of photons that arrive over the total period T. This technique requires a large number BT of detectors, and it is fortunate that other methods of setting up and detecting the signal are available. It can be shown that the basis set of functions in the positive frequency half of the expansion (8.14) need not be harmonic functions. Instead any set of BT orthonormal positive-frequency functions with period T may be used (subject to the restriction that the effective bandwidth is very much less than the mean frequency). To each such function there corresponds a mode of the loop and it is possible in principle to process each such mode independently of the others. Thus the basis set may consist of non-overlapping quasi-rectangular pulses modulating $\exp(2\pi j f_C t)$. It is possible to count the number of photons in each pulse, by using a single detector and by using the time of arrival to determine to which pulse a photon belonged.

Alternatively as in Section 8.2.5 we may determine the in-phase and quadrature components of a given pulse, as measured relative to the carrier. In principle other orthogonal basis sets may be used, but the harmonic function basis and the pulse-function basis lead to the most obvious receiving techniques.

The old definition of the bandwidth equivalent B (Section 5.1.2) has to be generalized in the quantum domain because in quantum theory the state of a mode is not specified by a complex number. Therefore we shall redefine B as the number of normal modes to be measured as completely as possible in a signal of duration T, divided by T. (In the prequantum theory each complete measurement determines two real numbers or one complex number.) When the modes are represented by an orthogonal set of harmonic functions this quantity B is just the bandwidth. When they are represented by non-overlapping (and hence orthogonal) pulse-functions, we find that B is just the pulse-rate. Thus the use of B as the pulse-rate in Section 8.2.4 is completely consistent with this new definition. Similarly $\bar{n}$ can be generalized to denote the mean number of photons per mode. In fact this symbol has already been used to denote the number of photons per pulse in Section 8.2.4, and the number of photons per normal mode in this section.

8.3.5 Comparisons with the upper bound

Finally we show that the idealized detector-systems come quite close to the upper limit (8.16). This means that no other system will be found to perform significantly better. The basic parameter $\bar{n}$ is the ratio of the photon rate to the bandwidth equivalent B, and is small in wideband systems and large in narrowband systems. We consider these situations in turn:

(a) $\bar{n} \ll 1$. The upper bound (8.16) gives

$$J'_{UB}/B = \bar{n} \ln \bar{n}^{-1} + \bar{n} + O(\bar{n}^2).$$

The incoherent system of Section 8.2.4 achieves

$$J'/B \approx 0.673\, \bar{n} \ln \bar{n}^{-1},$$

that is about $\tfrac{2}{3}$ of the upper bound. The coherent system of Section 8.2.5 has a Shannon limit of

$$J'/B = \ln(1+\bar{n}) = \bar{n} + O(\bar{n}^2)$$

(with the quantum efficiency η_Q set to 1 for ideal detection). This lacks the factor $\ln \bar{n}^{-1}$. So incoherent detection is superior for $\bar{n} \ll 1$, and is reasonably close to the quantum bound.

(b) $\bar{n} \gg 1$. This time we obtain

$$J'_{UB} = \ln \bar{n} + 1 + O(\bar{n}^{-1}) = \ln(e\bar{n}) + O(\bar{n}^{-1})$$

from (8.16). The coherent system has a Shannon limit of

$$J'/B = \ln(1 + \bar{n}) = \ln \bar{n} + O(\bar{n}^{-1}),$$

which at least contains the logarithmic leading term, but needs a factor $e = 2.718$ increase in energy to match the upper bound, i.e. 4.3 dB. The incoherent system receives the signal at B bits per second with negligible error probability, and so has a scaled rate $J' = B \ln 2$ which gives

$$J/B = 1.$$

It is an inferior system, not only because it fails to use the signal-phase to convey information, but also because it uses only two amplitude-levels.

In the intermediate region ($\bar{n} \approx 1$) both the coherent and incoherent system compare reasonably with the upper bound.

Exercises

8.1 Compute the maximum possible current through a photodiode due to 1 μW of radiation at 300 THz.

8.2 Assuming that a sixth magnitude star is just visible, estimate the lowest photon-rate to which the eye is sensitive. (Take area of pupil as 20 mm^2, and frequency of radiation as 500 THz.)

8.3 A fibre-optic link operating at 0.85 μm achieves an information rate of 800 M bits/s with a reliability of 10^{-9} at a power level of -35.5 dBm at the detector. What is the mean number of photons per bit of information?

8.4 In the incoherent system of photodetection and amplification described in Section 8.2.2 the input capacity of the amplifier was taken as 1 pF. Assuming that the minimum value of this capacity is determined by the detector as the capacity of a capacitor of side one wavelength (in vacuum), i.e. by $\epsilon_0 \lambda$, where λ is the wavelength and ϵ_0 ($= 8.854 \times 10^{-12}$ C V^{-1} m^{-1}) is the permittivity of free-space, recompute the estimate $N \geqslant 750$ for the number of photons needed in a pulse.

8.5 Use the methods of Section 8.2.5 to find the mean and variance of the number of detected photons if $\lambda(t) = \lambda$, a constant, and if $C(t) = 1$.

8.6 Derive the Poisson distribution (8.4) using the device of breaking up a time-interval into infinitesimal time-intervals during each of which a photon may arrive with a probability independent of what happens in the other intervals.

8.7 Optimize the information rate of the photon-detection system (Section 8.2.4) with respect to Q.

8.8 Calculate the mean photon number for a normal mode at 600 THz for $\theta_N = 300$ K, 3000 K.

8.9 Estimate the rate of transmission of power down a transmission line due to a source at temperature θ at one end.

8.10 Estimate the maximum information rate through a lens of area A when power P is flowing into it. (Assume that the lens acts as A/λ^2 separate channels where λ is the wavelength of radiation at a typical frequency.)

8.11 Estimate the maximum information rate possible with a narrow-band signal of bandwidth 1 GHz at a carrier-frequency of 500 THz, using a power of 1 W at a distance of 10^{12} m, with ideal mirrors of effective area 1 m^2 at the transmitter and receiver.

9. Cipher systems

9.1 Introduction

9.1.1 *Basic terminology*

In this final chapter a different aspect of communications theory is considered, the question of security rather than reliability. The widespread use of electronic communications in a commercial environment means that a great deal of data which was sent in a fairly secure manner in the past is now sent by communications links to which many people potentially have access, authorized or unauthorized. This is particularly true of microwave and satellite links. Hence there is a need for techniques for concealing the contents of a message and for detecting any tampering with a message. The most important technique is the use of special codes known as *ciphers*. Ciphers have been used by the armed forces and by diplomatic services for centuries, but their use on a wide scale in a commercial context is fairly recent, a consequence of the methods of modern communications. At the same time intriguing new concepts and types of cipher have been discovered. Some of these new ideas will be discussed in this chapter.

A quick run-through of the terminology is in order at this stage. We suppose that one person, the *sender*, wishes to send another person, the *recipient*, a message which he wants to keep secret from an *eavesdropper*. The message must be transmitted over an insecure channel, to which it must be presumed the eavesdropper has access. The message is called the *plaintext*. It is *enciphered* or *encrypted* by an algorithm or set of rules called the *encryption algorithm*. This algorithm is controlled by a string of symbols called the *key*. The key is kept secret from everyone except the sender and recipient, and it should be easily changed in case it has somehow been discovered by the eavesdropper. The output from this algorithm is called the *ciphertext*. The key is also known to the recipient who uses it in an algorithm inverse to the sender's algorithm to recreate from the key and from the ciphertext the original plaintext. This is called *deciphering* or *decryption*. The eavesdropper is presumed to have copied the ciphertext from the channel and is presumed to know the encryption and decryption algorithms, without knowing the key. The breaking of a cipher system by an eavesdropper without the use of the key is called *cryptanalysis*.

Other reasons for using ciphers are to solve the problems of *authentication* and of *disputes*. The problem of authentication is how to make sure that the person you are communicating with is the person you think you are communicating with, and not some impostor. Conventional cipher systems usually provide a solution automatically, since in most cases only two parties know (or should know) how to communicate with a given cipher-system. The problem of disputes is the following. Suppose A sends B a message (which may or may not be encrypted). Later A denies that he ever sent it and accuses B of making it all up. The conventional way of preventing this problem is to use a *signature*. B compels A to sign any message A sends him with an unforgeable signature, by refusing to take any notice of unsigned messages. In cases where this does not provide sufficient security a third party trusted by both A and B is called in as a witness to the communication. Both these solutions are rather clumsy in electronic systems.

Modern trends are towards public algorithms for encryption and decryption, at least in commercial environments, since these algorithms are often computer programs which are readily copied surreptitiously. Thus the assumption that no one else knows about them is dangerous. In fact in 1976 a complete specification of an algorithm was published, known as the Data Encryption Standard (Section 9.4.3). Only the key is kept secret. The ultimate step in the direction of openness came with the public-key ciphers where even the encryption key is published.

In traditional systems the management of keys becomes a serious problem once there are a large number of users, since a distinct key must be assigned to each pair of users. Moreover it is not easy for two people, who have just found out that they have an interest in common and who wish to get into communication, to start using encryption immediately, since somehow a key must be sent from one to the other by secure means.

9.1.2 *Monoalphabetic and bigram ciphers*

Some old-fashioned ciphers are now considered to illustrate the principles of cryptography. In a *monoalphabetic* cipher the plaintext is usually written in capitals without punctuation or spaces between the words. (The 26-letter alphabet with capitals only will be referred to as the Roman alphabet.) An obvious cipher is to use a special character for each letter (Conan Doyle 1981). Such a transliteration cannot be sent by a telegraph link and so instead one simply replaces each letter by another letter of the alphabet, so that the key is effectively a permutation on 26 letters. There are $26! = 4 \times 10^{26}$ possibilities and to break the cipher by trying each one out in turn is out of the question. Testing each one every microsecond would take over 10^{13} years, almost a thousand times the age of the universe as currently postulated. But the method used by Sherlock Holmes works rapidly enough (Conan Doyle 1981). Some letters of the alphabet tend to occur much more frequently than the others. Since E is likely to be the commonest letter in the plaintext, the commonest letter in the cipher most likely stands for E. After the commonest letters have been assigned one searches for common groupings of letters in the plaintext like 'the' and 'tion', or for likely words if we know the context of the message.

In a bigram cipher the letters in a message are paired off and each pair is represented by another pair. The substitution table contains $26^2 = 676$ entries and so is rather hard to keep to hand, but it is possible to set up a system which uses a short key to produce a rule equivalent to using such a table. Such a cipher is harder to break than the monoalphabetic cipher, but a table of frequences of letter-pairs may be used to spot the common pairings. Thus in English plaintext TH, HE and ER are frequent pairs, and their representations in the cipher text are soon found.

9.1.3 *Running-key cipher*

In a running-key cipher the key specifies a text of which identical copies are available to sender and recipient. As before all the letters are capitalized and spaces and punctuation are removed both from the plaintext and the keytext. The letters are then regarded as integers from 0 to 25 with $A = 0$ and $Z = 25$. Corresponding letters from the plaintext and the specified text are then added modulo 26. Subtraction is an alternative if this has been agreed to in advance.

9.1.4 *Some further comments*

The ciphers described so far are simple precursors of types of cipher in modern use. The monoalphabetic and bigram ciphers are examples of *block ciphers* where each block (of one and two characters respectively) is enciphered into a similar block, independently of what happens to the other blocks. The running-key cipher is a typical *stream cipher* where the key is used to generate a keystream which is then added in to the plaintext (using modular arithmetic). For convenience in digital processing their modern successors normally use a binary alphabet with a size equal to a power of 2.

Another traditional cipher not so far described is the polyalphabetic cipher. Here several different substitution tables are used in turn. Thus suppose that the chosen number of tables is 6. Then the 1st, 7th, 13th, . . . characters are enciphered by the first table, the 2nd 8th, 14th, . . . by the second and so on. The number of tables is called the period, for obvious reasons. If the eavesdropper knows the period, 6 in this case, he first sorts the ciphertext characters into 6 sets with the characters numbered 1, 7, 13, . . . in the first, 2, 8, 14, . . . in the second and so on. Then frequency analysis is applied to each set in turn, as though it were a monoalphabetic substitution cipher. If the eavesdropper does not know the period then he tries different values until the highly non-uniform character-frequency distribution of a monoalphabetic substitution shows up. The weakness of this cipher is that although the key is quite large, being in effect 6 substitution tables, only one table is used for each character.

Major causes of the weakness in the monoalphabetic and polyalphabetic ciphers are that only a small part of the key is used for each encipherment, and that a small change in the key affects only a small part of the ciphertext.

Thus if two letters in the substitution table are exchanged then the ciphertext is not much altered. As parts of the message begin to make sense the eavesdropper can tell that his ideas about the key are not far from the truth. To avoid this problem modern computerized ciphers arrange that even if just one bit of the key is altered every part of the ciphertext is liable to be changed.

9.1.5 *Functions and mappings*

A specified relation between a set of 'input values' and a set of 'output values' may be regarded as a function, a table of values, an algorithm, a mapping, or a transform. Sometimes one point of view is preferable to another. When we regard the relation as a function we use a symbol like f for the function and write $f(x)$ for the output specified by the input x. We may also regard the relation as specified by a lookup table, a table of values assigning an output to each input. We may also regard the relation as a mapping, where the inputs are represented by points in one set, the 'input set', the outputs as points in another set, the 'output set', and the relation is shown by a directed line (a line with an arrow) joining each input to its corresponding output. (Every input must be joined to some output, but we allow points in the output set which are not joined to any input. The output set is the set of conceivable outputs, not all of which may occur in practice.) The relation may also be regarded as an algorithm or even a computer program which is a rule for generating an output for every possible input. Sometimes we talk about the output as being a transform of the input.

An example is a code like the ASCII code where each symbol from a set of symbols is associated with a seven-bit string. Here the output is of a different kind from the input.

Another example is provided by the monoalphabetic substitution cipher. Here the set of inputs is the Roman alphabet, as is also the set of outputs, and the relation may be specified by a lookup table. In some cases the relation is more conveniently specified by a rule or algorithm such as: 'Number the letters from 0 for A to 25 for Z. Take the sum "modulo 26" of 5 and the number of the input, and give as output the letter numbered by this sum.' Thus with this relation the outputs corresponding to B and Y are respectively G and D, and if we regard the relation as a function with the symbol f we may write $G = f(B)$, $Y = f(D)$.

Next we need to discuss the inverse of a function. The relation is best regarded as a mapping. To obtain the inverse we reverse the arrows and interchange the names 'input set' and 'output set'. For this to work we must ensure the every output in the original mapping corresponds to at most one input, for otherwise in the new mapping we shall have one input with more than one output. If this happens we say that the relation has no inverse. If there are unused outputs in the original mapping then again we shall break a rule by having inputs in the new mapping with no output, but this problem can usually be overcome by joining these unattached inputs to a newly created output, denoted by say '?'. For example in the above substitution cipher we may write the inverse relation as f^{-1} and we then write $f^{-1}(Y) = D$. Evidently in this case every input is in one-to-one correspondence with every output, and vice-versa.

In a cipher the relation of output to input, commonly called a cryptographic transform, is usually selected by a key k from a set of keys. There are two ways of looking at this situation. In the first way we regard the key as choosing one function out of a possible set. We then label the function by the symbol for the key, so that the output corresponding to the input x when the key is k may be written as $f_k(x)$. Alternatively we may picture the input x and the key k being fed together into the appropriate inputs of some algorithm. In this case we say that the output depends on the pair (x, k) and we may imagine it as given by a function with two inputs: $f(x, k)$. Thus suppose that in the above substitution the key is the value of the shift, which happens to be 5. Then we could write either $f_5(B) = G$ or $f(B, 5) = G$ depending on which notation is better in the context.

When the input and output sets are identical and of finite size and when each input is associated with one output and vice-versa, then the mapping is called a 'permutation'. Of course it is always invertible. The number of permutations on n symbols, that is the number of invertible mappings of the set of n symbols into itself, is given by

$$n! = n(n-1)(n-2)\ldots 3.2.1$$

as is shown in any text on elementary probability theory.

9.2 Conventional ciphers

9.2.1 *The one-time pad*

Suppose we wish to send an N-letter plaintext **p** and we choose completely at random an N-letter string **k** which is then added letter-by-letter to **p** to give the cipher-text **c**:

$$\mathbf{c} = \mathbf{k} + \mathbf{p}. \tag{9.1}$$

(The addition is done as described in the last section.) The official recipient of this message has also been given a copy of **k** by some secure route, and so he re-obtains **p** by subtracting **k** from **c**:

$$\mathbf{p} = \mathbf{c} - \mathbf{k}. \tag{9.2}$$

The eavesdropper knows **c** without knowing **k**. It turns out that the only thing he can find out about **p** is its length N. He could interpret **c** according to (9.1) but also as

$$\mathbf{c} = \mathbf{k}' + \mathbf{p}' \tag{9.3}$$

where $\mathbf{p}'$ is some other reasonable message of N letters. Thus $\mathbf{k}'$ is given by

$$\mathbf{k}' = \mathbf{k} + \mathbf{p} - \mathbf{p}'. \tag{9.4}$$

But this other N-string $\mathbf{k}'$ is just as likely to have been added to the plaintext message as $\mathbf{k}$ and so $\mathbf{p}'$ is just as good a decipherment as $\mathbf{p}$. This is true for $\mathbf{p}'$ equal to any reasonable plaintext. The eavesdropper's problem is not that he cannot find a plaintext; it is that he can find too many possibilities and will not know which to choose.

This technique can be used even with $N = 1$ and with an alphabet of size 2 when there is just one bit of information to be sent. The sender and receiver have previously agreed on a one-bit key, say by tossing a coin. This key is then added by the sender to the one-bit message and subtracted again by the receiver. (Of course in the binary case addition is the same as subtraction.) An eavesdropper then has a fifty-fifty chance of being right about the transmitted message, but he would still have this probability if he just tossed a coin without bothering to listen in.

In most commercial environments messages are much longer than one bit. The problem then arises that the keystring, which is just as long as the message, has to be sent in advance by a secure route from the sender to the receiver (or vice-versa). The obvious temptation is to keep reusing the same keystring, but as will soon be explained this is a terrible mistake. Perhaps as a reminder of this the system is called the 'one-time pad' or 'one-time tape' system, depending on the medium on which the key-string is written. The system is thus restricted to special applications where (a) hopefully there is not much traffic, (b) a complete guarantee of privacy is required, and (c) economic considerations are secondary. For instance something like this might be used on the Moscow–Washington 'hotline'.

9.2.2 Theoretical secrecy

Shannon explored these ideas further and developed the concept of 'theoretical secrecy' where the eavesdropper is unable to decide on the plaintext because there are too many possibilities. With an alphabet of size q there are $C = q^N$ N-symbol strings. Let us assume that there are also $P = r^N$ intelligible English plaintexts of length N which are all equally likely (Section 3.1.1). Here r is considerably smaller than q. Estimates for $\log_2 r$ range from 1.0 to 1.5 (Beker and Piper 1982, Secton 3.6.3), definitely smaller than $\log_2 q = 4.7$ when $q = 26$. Finally let K be the number of possible keys $\mathbf{k}$. The encryption process is an algorithm which accepts as input any possible key $\mathbf{k}$ and any possible plaintext $\mathbf{p}$ and gives a ciphertext $\mathbf{c}$ as output. The input may be written as an ordered pair $(\mathbf{k}, \mathbf{p})$. There are altogether KP such pairs, an astronomically large number of course in any practical situation. We shall imagine that this algorithm has good randomizing properties so that without deep study all we can say without knowing **k** and **p** is that any randomly chosen N-string is as likely to be the output as any other. If $C \gg KP$ then it is very unlikely that two different input-pairs $(\mathbf{k}, \mathbf{p})$ and $(\mathbf{k}', \mathbf{p}')$ give rise to the same output. The chances that a given N-string is formed at all as we try all possible KP input pairs is of order $(KP)/C$ and so if it is formed once it is very unlikely that it will be formed by another $(\mathbf{k}, \mathbf{p})$ input-pair. On the other hand if $C \ll KP$, then every output string is probably formed many times (of order KP/C times) as we try all KP inputs $(\mathbf{k}, \mathbf{p})$. (The case $C \approx KP$ need not be considered since C and KP are huge numbers which are not likely to have anything like the same magnitude.) Thus in the first case a unique decipherment is usually possible since a ciphertext **c** can have come from only one input-pair. In the second case a ciphertext could have come from a very large number of possible input-pairs and it will not be possible to decide which one was actually used. Thus we find 'theoretical security' when $C \ll KP$, or taking logarithms, when

$$N \log_2 q < \log_2 K + N \log_2 r.$$

(Note that taking logarithms tends to mollify the fierceness of the inequalities. Thus we may safely assume that $10^{200} \gg 10^{100}$, but we could hardly write $200 \gg 100$.)

For a fixed-length key with K independent of N we find that this inequality is satisfied if N is small enough, in fact when $N < N_0$ with

$$N_0 = (\log_2 K)/(\log_2 q - \log_2 r).$$

As we might expect short texts are not uniquely decipherable, but this formula goes further by estimating the length N_0 of ciphertext below which unique decipherment is not possible. N_0 is known as the 'unicity distance'.

This is what happens for a fixed-length key. Let us now suppose that the key is as long as the plaintext and that the number of keys of length N is given by

$$K \approx l^N.$$

Then theoretical secrecy is possible if

$$\log_2 l > \log_2 q - \log_2 r.$$

In the one-time pad system l is equal to q and this condition is always satisfied.

In the running-key cipher the effective key is chosen as an English plaintext from a text known to both sender and receiver. The mod-26 sum of the plaintext and keytext is sent as the cipher. Thus the key to be sent by a secure channel is a brief reference. In this case $l = r$ and since $\log_2 q = 4.7$, with $\log_2 r$ in the range 1 to 1.5, we see that there is no theoretical secrecy. In fact there is not much practical secrecy either and it is found that the sum or difference of two English texts is readily taken apart. This is not done by searching all possible novels, encyclopaedias, training manuals, newspapers etc. for the key-text. A better way is to subtract (or add) mod-26 a likely word in all possible positions. If it matches, a piece of clear text will appear. Then one tries to extend this clear patch in both directions. If the match was a fluke then this will fail and one has to return to the first phase again. But otherwise the two texts can soon be teased apart.

Diffie and Hellman (1979) have suggested that theoretical security may be achieved by adding in several keytexts. Thus with four keytexts the condition for theoretical secrecy becomes $4 \log_2 r > \log_2 q - \log_2 r$ or $5 \log_2 r > \log_2 q$, which is satisfied if $\log_2 r \approx 1$, $\log_2 q \approx 4.7$. However such ciphers are not very practical in an electronic system since they require the storage of vast amounts of text.

We may now see why a one-time pad must not be used twice. Suppose we have two ciphertexts c_A and c_B formed by adding in the same key:

$$c_A = k + p_A, \quad c_B = k + p_B.$$

Then subtracting we find $c_A - c_B = p_A - p_B$, so that the difference is just a difference of two plaintexts, in fact a running key cipher in which one of the plaintexts acts as the key. This is readily broken and once p_A and p_B are known the eavesdropper immediately finds k, and is ready for any further messages using this key.

9.2.3 Known-plaintext attack

With keys of a fixed size and without the means for squeezing all redundancy out of a plaintext message we are left with the problem of finding cipher systems whose cryptanalysis takes up too much computer time or storage. At least we may hope that the time taken is so great that the value of the message to the eavesdropper has been lost by the time it is revealed. In order to sell a cipher system the designer has to persuade the prospective customer that even in the circumstances most favourable to the eavesdropper the eavesdropper cannot break the system. The usual circumstances considered are that the cryptanalyst knows the cipher text and a good part of the plaintext from which it originated and is determined to find the rest of the plaintext and the key if possible. The uninitiated reader may wonder why if the eavesdropper can get hold of the plaintext anyway anyone has bothered to use a cipher. But there are cases when this situation may arise:

(a) Commercial specifications or journalistic stories may be sent encrypted and then when everything is ready the plaintext is published.

(b) Material sent the previous time may have been cryptanalysed, and current messages may contain large chunks from the previously sent material. Alternatively the same message may have been sent twice by different ciphers, one of which has been broken.

(c) Many messages have a highly formalized structure and the cryptanalyst may well guess correctly a large part of the plaintext.

Older ciphers were vulnerable to the known-plaintext attack. This led to the absurd situation that plaintext messages had to be kept secret long after they had ceased to matter, just because they had been sent by cipher, and other plaintexts which were still sensitive had also been sent using the same key.

9.3 Stream ciphers

9.3.1 *Introductory*

From now on it will be supposed that the plaintexts and cipher texts are binary strings. The binary equivalent of the one-time pad is to produce the ciphertext using bit-by-bit addition (modulo 2) of a randomly generated binary *key-stream*, stored perhaps on a magnetic tape. Since the key-stream is as long as the plaintext such a cipher-system is usually impractical.

Instead one naturally considers using a 'pseudo-random' *key-stream* produced by an algorithm not unlike those used for generating 'random' numbers on a computer. The sequence of values produced by such an algorithm is completely determined once the initial setting is specified, the 'seed' as it is usually called in that context. None the less as judged by statistical tests the sequence produced by a good algorithm appears to be random. In the present application the secret key acts as the seed for such an algorithm, implemented at both the sending and receiving ends. Provided both sender and recipient have previously agreed on a key they can produce identical key-streams. Instead of the key filling a bulky magnetic tape it is now a binary string perhaps a hundred or so bits long. Ciphers where a key-stream is added bit by bit or character by character to the plaintext are known as *stream ciphers*.

9.3.2 *Feedback shift-registers*

One method of producing pseudo-random bit strings is the use of feedback shift registers like those described in Section 7.4. These are simple to implement, are very fast in operation, and can be made to generate statistically very good sequences, provided the registers are long enough. Figure 9.1 shows a four-bit shift register, much shorter of course than would be used in practice. In general such a shift register consists of m locations labelled from 0 on the left to $m-1$ on the right, which hold a binary m-string. The shift-register is stepped on by (a) shifting all the bits left one location, with a 0 coming into the right-hand location, (b) feeding back the old left-hand bit along the wire to be added in to the locations indicated, immediately after the shift. In algebraic terms this may be expressed as follows: If p_i, p_i' are respectively the contents of the ith location before and after the step, then

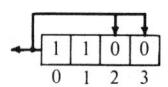

FIG. 9.1 Feedback shift-register for generating pseudo-random bit-strings.

$$p'_i = p_{i+1} + f_i p_0, \qquad i = 0, 1, \ldots, (m-1)$$

with p_m taken as zero. Here f_i equals 0 if there is no feedback connection into the ith location, and equals 1 if there is. In the example of Fig. 9.1 we have $m = 4$ and $f_0 = 0, f_1 = 0, f_2 = 1$, and $f_3 = 1$. With an initial setting of $(p_0, p_1, p_2, p_3) = (1, 1, 0, 0)$ we find the following successive states (i.e. values of (p_0, p_1, p_2, p_3)): $(1, 1, 0, 0), (1, 0, 1, 1), (0, 1, 0, 1), (1, 0, 1, 0), (0, 1, 1, 1)$ etc. The pattern repeats after 15 steps. The output sequence is (reading from left to right) 1, 1, 0, 1, 0, 1, 1, 1, 1, 0, 0, 0, 1, 0, 0, and then it repeats indefinitely.

An alternative method of setting up feedback shift-registers is described by Beker and Piper (1982, Section 5.3.2) where the flow is apparently the other way. It is not hard to show that the two arrangements can be made to produce the same output sequence. It is easier to prove a number of theorems about shift-register output sequences using the alternative arrangement. However we are not much concerned with these here.

It can be shown that the output sequence must eventually repeat with a period which is at most $2^m - 1$. If the feedback connections have been chosen to give this maximal period the output sequence depends on the initial contents only for its starting point (provided the initial contents are not the string of nulls). Thus with the initial contents 0101 instead of 1100 the new sequence starts at the third position of the old sequence. Since the output sequences can be shown to have good quasi-random properties, it seems reasonable to use them as key-streams, provided of course m is chosen large enough and provided that the feedback is chosen to give the maximal period $2^m - 1$. In this case the output sequence is known as a maximal-length feedback shift-register sequence or m-sequence for short. It is also sometimes called a pseudonoise or PN sequence to emphasize its apparently random properties. Thus we may encipher a message by adding in bit-by-bit the output string from a feedback shift register. The key would specify the feedback connections and the initial contents of the feedback shift register.

9.3.3 The division algorithm revisited

Unfortunately as it stands this cipher-system is very easily broken by a known plaintext attack. Knowledge of part of the plaintext together with the ciphertext enables the eavesdropper to determine immediately a portion of the m-squence used for the encryption. Provided he has at least $2m$ consecutive bits he has in principle enough information to determine the m feedback settings

$f_0, \ldots, f_{m-1}$ and the m initial connections $p_0, \ldots, p_{m-1}$ (at the start of the keystream sequence as obtained by the eavesdropper, rather than at the start of the transmission). In practice this is rather easily done. If the output sequence $k_0, k_1, \ldots, k_{d-1}$ up to d bits is represented as a polynomial $K(z) = k_0 + k_1 z + \ldots + k_{d-1} z^{d-1}$, then it is readily shown that $K(z)$ may be regarded as the ratio of the polynomials

$$P(z) = p_0 + p_1 z + \ldots + p_{m-1} z^{m-1}$$

and

$$Q(z) = 1 + z(f_0 + \ldots + f_{m-1} z^{m-1}),$$

taken modulo z^d. In algebraic terms we may write

$$P(z) \equiv K(z) Q(z) \bmod z^d, \tag{9.5}$$

This congruence is of the same form as the 'key equation' (7.40) used in the algebraic decoding of the codes described in Section 7.7. So the same technique may be used to solve it, provided that $d \geq 2m$. Thus this cipher is very readily broken.

The output polynomial K is related to P and Q by (9.5) because the shift register of Fig. 9.1 implements a division algorithm. A simple way to demonstrate this is to illustrate the algorithm in the case shown in Fig. 9.1 with $m = 4$, $Q(z) = 1 + z(z^2 + z^3) = 1 + z^3 + z^4$, and $P(z) = 1 + z$, corresponding to the initial setting $(1, 1, 0, 0)$. Note that the polynomials are written in increasing powers of z, unlike those used in the division algorithm of Section 7.3.3. This time multiples of $Q(z)$ of the form $z^l Q(z)$ are used to remove the *lowest* powers of z from $P(z)$. The constant term (with $l = 0$) is removed at the start. The process is usually written down as a long-division sum, using detached coefficients (Table 9.1). Every time a shifted version of Q is subtracted a '1' is entered in the row K representing the quotient. The result of each subtraction is a partial remainder R_l, labelled by the lowest power of z with a non-zero coefficient. Thus $R_1 = P - z^0 Q$, $R_3 = R_1 - z^1 Q$ and so on. K accumulates the subtracted powers of z (without the factor Q) so that after l stages K is the polynomial $K_{l-1}(z)$ of degree $(l-1)$ at most. It is evident that

$$P(z) = Q(z) K_{l-1}(z) + R_l(z)$$

from which (9.5) readily follows, since the process stops when $l = d$.

It is also evident that this is the long division implemented by the shift register. The feedback implements the subtraction process. The only slight difference is that any stage involving the subtraction of zero is left out of the long division for brevity, although of course it still contributes a bit 0 to the output string K.

TABLE 9.1

	z^0	z^1	z^2	z^3	z^4	z^5	z^6	z^7	z^8	z^9	
Q:	1	0	0	1	1						
K:	1	1	0	1	0	1					
P:	1	1	0	0							
	1	0	0	1	1						
R_1:		1	0	1	1						
		1	0	0	1	1					
R_3:				1	0	1					
				1	0	0	1	1			
R_5:						1	1	1			
						1	0	0	1	1	
R_6:							1	1	1	1	etc.

9.3.4 Introducing non-linearity

The use of non-linear feedback in the shift registers is an obvious remedy, but unfortunately the theory is not well developed, and there is little guarantee that the output sequence will have good statistical properties. For instance a sequence where the number of 0s far outweighs the number of 1s gives very little concealment. So instead a non-linear function of the output from one or more linear-feedback shift registers is used. Here is an example of what is possible. (These systems have not been analysed by the author and he refuses to guarantee their security.)

A shift register or its output sequence is tapped at p fixed locations spaced irregularly. Let the binary value at the ith tap be a_i. Then the p tapped values, $a_{p-1}, a_{p-2}, \ldots, a_1, a_0$ specify an integer L in the range 0 to $2^p - 1$ through its binary expression:

$$L = 2^{p-1} a_{p-1} + 2^{p-2} a_{p-2} + \ldots + 2a_1 + a_0.$$

Then the Lth bit in a list of 2^p bits labelled from 0 to $2^p - 1$ is chosen for the output. The list may consist of a selection of bits chosen at random, or of part of the binary expansion of some irrational number such as π or e whose statistical properties have been well studied. To implement this system the p taps are used as the address lines of a computer-type memory holding one bit in each address.

A variation on this idea is to use a feedback shift register with a length of at least 2^p instead of the fixed list. The Lth location of this register is chosen to provide the output bit, after which this register is stepped at the same time as the other which picks the address. (Beker and Piper 1982 call such a system a 'multiplexed' system, and give some useful theoretical results for it.) Another possibility is to use p distinct registers to drive the address lines. The periods of these registers should be coprime to ensure the maximum period in the sequence of address values L and thus hopefully to maximize the period of the output sequence.

These systems may sound effective, but they should if possible be subjected to mathematical analysis, or at the very least to public scrutiny to see if they can be broken. If this is not possible then the user will always be left with a slight doubt about the security of the system, perhaps exacerbated by the eavesdropper who hopes to persuade the legitimate user to change to a less secure system. Unfortunately this is true for most systems, with the notable exception of the one-time pad.

9.3.5 *Other problems*

(a) Stream ciphers have the advantage that if transmission errors cause a bit to be changed in the ciphertext then only the corresponding bit in the plaintext is changed. However if synchronization is lost during a fade, say, then it is not regained automatically and some form of synchronizing signal may have to be sent along with the ciphertext.

(b) If a legitimate third party wishes to join the communication part way through, then although he knows the initial settings of the shift registers (from the key) he has to arrange them to 'catch up' quickly.

(c) The same key may have to be used to send several different messages at various times. It is vital that the shift-registers are not initialized to the same value each time, for then we are effectively reusing the same key-stream over and over again, and this is totally insecure, for the same reason that a one-time tape should not be reused (Section 9.2.2).

(d) If the eavesdropper knows a part $\mathbf{p}$ of the plaintext as well as the corresponding part $\mathbf{c}$ of the ciphertext, and if he can tamper with the ciphertext before it is received by the legitimate receiver, then he can in effect replace $\mathbf{p}$ by another string $\mathbf{p}'$ more to his advantage. He does this by replacing $\mathbf{c}$ by $\mathbf{c}' = \mathbf{c} + \mathbf{p}' - \mathbf{p}$. Then the legitimate receiver computes $\mathbf{c}' - \mathbf{k} = (\mathbf{p} + \mathbf{k}) + \mathbf{p}' - \mathbf{p} - \mathbf{k} = \mathbf{p}'$. A highly formatted message is particularly susceptible to this sort of tampering. The eavesdropper could for instance replace one currency symbol by another, say $ by £ if he knows that this symbol always appears in the same place. To avoid this possibility some authentication system must be used, such as a check-sum. This checksum would have to be enciphered to prevent its being tampered with. This solution increases both the amount of computation needed and the length of the message to be sent.

One solution to the problems (a), (b), and (c) is along the following lines: The plaintext and hence the ciphertext are broken into blocks. Each block of ciphertext is transmitted after a synchronizing block. At the same time the system employs three keys, a *base key*, a *message key*, and an actual key. The base key is kept secret, and is fixed throughout the whole transmission. It is combined in some known manner with the message key to give the actual key for the enciphering algorithm. The message key is changed for each block and need not be kept secret. Hence it can be transmitted in the preceding synchronization block. Evidently this solves the problem of synchronization, and it makes it almost certain that the same actual key is not used twice. It also enables the legitimate late-arrival who knows the key to join the conversation after he has picked up the message key from the next synchronization block. The message-keys can be generated according to some fixed plan by a pseudo-random number generator say, and in this case it does not matter if the message-key is erased occasionally by noise, because it can be recreated from the plan.

9.4 Block ciphers

9.4.1 *Affine transforms*

An alternative system of encryption is to break the plaintext up into n-bit blocks, which are then 'scrambled' under the control of a key. This scrambling must be done in an invertible way so that the transform is a permutation on a set of 2^n symbols, each symbol being an n-bit string. Typical values of n are 64 and 128, such powers of 2 being very convenient for digital processing. (If each block is the binary representation of a single character, then such a cipher is a mono-alphabetic cipher.) An obvious scheme is to permute the bits and then change some of the bits, both operations being done under the control of the key. Such a transform is evidently easily inverted if the key and hence the permutation and the locations of the changes are known. Such a transform is a special case of an *affine* transform:

$$\mathbf{c} = \mathbf{Ap} + \mathbf{b}$$

where $\mathbf{p}$ is an n-bit block of plaintext regarded as a column vector, $\mathbf{c}$ is the corresponding block of ciphertext, $\mathbf{A}$ is a non-singular $n \times n$ binary matrix and $\mathbf{b}$ a binary vector whose set bits determine the locations to be changed. Here all the arithmetic is 'modulo 2'. Any permutation of the components of a vector can be effected by a suitably chosen matrix, so the transform just suggested is a special case. The word 'affine' simply means 'linear plus constant'.

Unfortunately the linearity of such transforms makes them very weak as cipher systems when subjected to a known-plaintext attack. The situation is somewhat akin to that for linear feedback shift registers. To see why, we start by considering a block cipher system using just a linear transform of each plaintext block $\mathbf{p}_i$ to give a ciphertext $\mathbf{c}_i = \mathbf{Ap}_i$. Here $\mathbf{A}$ is a non-singular matrix determined by the key. We now suppose that the eavesdropper knows many examples of plaintext blocks and their corresponding ciphertext blocks, although he does not know **A**. Suppose moreover he can find a set of n of these plaintext blocks which when put together as columns form a non-singular matrix **P**. Thus **P** may be the matrix whose columns are the first n blocks $\mathbf{p}_1, \ldots, \mathbf{p}_n$, although some of these n columns will have to be replaced by later plaintext blocks if **P** is singular. Let **C** be the matrix made up with the corresponding ciphertext blocks as columns. Then by the rules of matrix algebra we have

$$\mathbf{C} = \mathbf{AP}$$

and since **P** is invertible

$$\mathbf{A} = \mathbf{CP}^{-1}.$$

Such a computation is very easily carried out (even if n is of the order of tens of thousands) since in binary arithmetic there are no rounding errors or overflows. It is easy to find a non-singular matrix **P** since it can be shown that for a moderately large value of n the fraction of all $n \times n$ binary matrices that are non-singular is just over 28% (Exercise 9.5). So the chances of quickly finding a set of suitable plaintext blocks is good.

The eavesdropper does not have much more trouble with the more general affine transforms. He simply subtracts one of the plaintext-ciphertext pairs, the first say, from all the others. Then with

$$\mathbf{c}_i = \mathbf{A}\mathbf{p}_i + \mathbf{b}$$

it is found that

$$(\mathbf{c}_i - \mathbf{c}_1) = \mathbf{A}(\mathbf{p}_i - \mathbf{p}_1)$$

so that the unknown **b** has disappeared and the relation is linear. Once **A** is discovered **b** may be found by $\mathbf{b} = \mathbf{c}_1 - \mathbf{A}\mathbf{p}_1$.

9.4.2 *IBM systems*

In this section and the next we consider two families of block-cipher systems. The first family consists of systems developed by IBM (Feistel 1973). A member of the other family, to be considered in the next section, has been chosen as a standard cipher.

The IBM ciphers are based on 'S-boxes' (S for substitution) which typically transform four-bit strings into four-bit strings in an invertible manner. They can therefore be considered as permutations on an alphabet of sixteen symbols, each symbol being a four-bit string. There are altogether 16! such transforms of which only a small fraction are affine, although transforms which are 'nearly affine' must also be avoided. The transform is small enough to be implemented easily by a lookup table, and the inverse of any transform is readily found. These basic transforms are then built up into a multistage transform of a 128-bit block into a 128-bit block. Each stage consists of two parts: (a) The 128-bit block is broken into 32 four-bit 'groups' which are then transformed by S-boxes operating in parallel. (b) The output is then permuted by a fixed (unkeyed) permutation on all the 128 bits. This permutation is not intended to provide secrecy directly; since it is not set by the key it must be presumed to be known to a cunning eavesdropper. Its purpose is to make sure that the overall transform does not degenerate into thirty-two separate parallel four-bit transforms, when all security would be lost. Another way of describing the effect of these permutations is as follows: Suppose that a single bit in the input block is changed. The effect of this change should spread laterally until every bit in the output block is liable to be changed. A good cipher shows this error-propagation or avalanche effect, since it is necessary for security. About one-half of the bits in the output block are changed, since two apparently unrelated blocks would differ in about half the number of places, there being a fifty-fifty chance of two corresponding bits in unrelated blocks being different.

In a typical system there are sixteen stages. Overall this needs $32 \times 16 = 512$ S-boxes. To avoid a key of excessive length, each S-box is chosen from just two possibilities, so that only 512 bits of key are needed. Even this is rather lengthy and so each of these 512 bits is generated from a shorter key. This is done in such a way that the change of a single bit in the key is liable to change every S-box. We note a few further points:

(a) If the S-boxes are affine then the overall transform is also affine, and may be broken by a known-plaintext attack as described in Section 9.4.1.

(b) Each stage should carry out a different transform from the others. Otherwise the system is susceptible to a chosen plaintext attack, of a sort described by Beker and Piper (1982, Section 7.3.3).

(c) The inverse process of deciphering involves passing each block backwards through each stage, undoing every transform. This is evidently just as simple as the enciphering process, although it is not in general identical to it.

(d) For high-speed use, say on the tape-drives of a computer, it is very inefficient to wait for one block of plaintext to pass through all the stages before the next one is loaded in. A following block of data can be loaded into a given stage as soon as that stage has passed on the previous block. (This method of using the processor is known as 'pipelining'. This is by analogy with a real pipe, where one need not wait for one drop of fluid to travel the full length before the next one goes in.)

(e) The effects of errors in transmission are severe. A good block cipher produces a completely different ciphertext if just one bit of the plaintext is changed, and a similar phenomenon occurs in the inverse transform. Thus a single bit in error in a cipher block completely garbles the plaintext. So if the transmission channel is noisy some form of error-correction will have to be used on the ciphertext, with rather greater power than would normally be

BLOCK CIPHERS

needed for sending plaintext. Thus the overheads are increased. On the other hand the cipher can be used for error-detection: if each plaintext block contains some sequencing information, then if this is garbled or out of order some form of error is noted. Loss of synchronization is not so serious as in stream-cipher systems, since each block starts from scratch, as it were. For the same reason it is very easy for a legitimate third party to join in late.

9.4.3 *Feistel cipher and the DES system*

A slightly different family of block ciphers is based on the following idea, the Feistel cipher (Beker and Piper 1982). As in the IBM systems the blocks are passed through several stages. The blocks must have an even length n and at each stage the input is broken into two equal left and right halves l and $\mathbf{r}$. the corresponding halves of the output are given by

$$l' = \mathbf{r} + \mathbf{u}(\mathbf{k}_i, l), \qquad \mathbf{r}' = l.$$

where $\mathbf{u}(\mathbf{k}_i, l)$ is an enciphered version (of length $n/2$) of the string l, with the transform controlled by a subkey $\mathbf{k}_i$ for the ith stage. This subkey is formed from the key $\mathbf{k}$ in a way which makes it depend on the whole of $\mathbf{k}$, and it should vary from stage to stage. Evidently only one half has been transformed by this stage; the other half is transformed at the next stage. An important feature of this system is that the function $\mathbf{u}$ does not have to be inverted to achieve the deciphering. In order to derive l and $\mathbf{r}$ from l' and $\mathbf{r}'$ we note that

$$l = \mathbf{r}', \qquad \mathbf{r} = l' - \mathbf{u}(\mathbf{k}_i, l) = l' - \mathbf{u}(\mathbf{k}_i, \mathbf{r}').$$

Thus the deciphering process operates in reverse order, but the function $\mathbf{u}$ is the same as in the enciphering process. Since the inverse of $\mathbf{u}$ does not have to be found and need not even exist we have considerable freedom in choosing $\mathbf{u}$.

The Data Encryption Standard (DES) is based on this system (National Bureau of Standards, 1977). The block length is sixty-four bits and there are sixteen stages. A full description may be found in Beker and Piper (1982, Chapter 7). We note the following points about this cipher:

(a) It is the first time that a complete algorithm has been published as a standard. In consequence the hardware for implementing this system is mass-produced since it is in fair demand. On the other hand its universal application attracts the attention of ingenious cryptanalysts, who may find a weakness in it. Whether such a discovery is published is another matter.

(b) The key for the DES is fifty-six bits long. With a known plaintext an exhaustive search requires $2^{56} = 7 \times 10^{16}$ trials. With a million chips operating in parallel, each taking a microsecond for each trial, the overall search takes about a day. Since the hardware is mass-produced such an arrangement may not be beyond the means of a large corporation (Diffie and Hellman 1979).

It may well be that this standard will be replaced in the next few years by a cipher with a larger key. On the other hand if this is the only weakness of the cipher, then it may well be incorporated as a building block in a larger system using a larger key (Meyer and Matyas 1982).

9.4.4 Cipher feedback

Although an eavesdropper may not be able to decipher each block, he may be able to record a block and use it to replace another block in the cipher text, or to add it to the ciphertext. In order to detect such tampering the legitimate users must have some way of sequencing the blocks. Part of each plaintext block may contain a sequence number, but this may seriously curtail the space available for text. We now describe the *cipher feedback* and *block chaining* methods for chaining the blocks without significantly increasing the overheads (Beker and Piper 1982, Chaper 7).

The straightforward encipherment may be represented as the transformation of a sequence of plaintext blocks p_i into ciphertext blocks c_i:

$$c_i = f_k(p_i), \qquad (9.6)$$

where f_k is the transform determined by the key k. The decipherment is then the inverse transform:

$$p_i = f_k^{-1}(c_i). \qquad (9.7)$$

There are several possibilities for chaining the blocks together.

(a) *Cipher feedback* is the use of the algorithm described by the formula

$$c_i = p_i + f_k(c_{i-1}). \qquad (9.8)$$

The previous cipher block is encrypted and then added to the current block of plaintext. The inverse process is

$$p_i = c_i - f_k(c_{i-1}). \qquad (9.9)$$

We note that the function f_k need not be invertible. An error in receiving a ciphertext block c_{i-1} will cause an error in p_{i-1} and also in p_i, but after that the transmission proceeds normally. Moreover a legitimate third party may join in late without losing more than one block after he joins.

(b) A modified version is as follows: The plaintext and ciphertext are divided into 'small blocks', each of a length l, a fixed submultiple of the blocksize L of the encryption algorithm. Thus the number of small blocks in such a block is $m = L/l$. (The choices $l = 1$ or $l = L$ are permissible.) Then the encryption process is as follows:

$$p_i = c_i + g_k(c_{i-1}, c_{i-2}, \ldots, c_{i-m}) \qquad (9.10)$$

BLOCK CIPHERS

where the function $\mathbf{g}_k$ joins up the small blocks into a block of size L, encrypts that with the transform $\mathbf{f}_k$ and then picks out a specified small block from the output. The inverse is evidently

$$\mathbf{c}_i = \mathbf{p}_i - \mathbf{g}_k(\mathbf{c}_{i-1}, \mathbf{c}_{i-2}, \ldots, \mathbf{c}_{i-m}) \tag{9.11}$$

so that as before the inverse of the encryption function is not needed. One disadvantage of this process is that since all but a fraction $1/m$ of the output of the encryption algorithm is wasted this algorithm has to work at a rate m times as fast as in a system without feedback.

(c) 'Pipelining' is not possible if we have to wait for the output from the encryption process to come all the way through each time. An alternative to (9.8) is to use

$$\mathbf{c}_i = \mathbf{p}_i + \mathbf{f}_k(\mathbf{c}_{i-d}) \tag{9.12}$$

where d ($\geqslant 1$) is the delay. In this way it is arranged that the block from $\mathbf{f}_k$ needed at the ith stage has been fed in d stages earlier. (This may reduce the security, since adjacent blocks are not linked togther.)

(d) *Block chaining*: In a public-key cipher system the encryption algorithm is publicly known. (However the inverse transform is in practice unattainable knowing only the encryption transform.) In this case cipher feedback provides no secrecy since the inverse transform is not used in the decipherment (9.9). The following system (known as block chaining) may be used instead: We set

$$\mathbf{c}_i = \mathbf{f}_k(\mathbf{p}_i + \mathbf{c}_{i-1}) \tag{9.13}$$

so that the previous cipher block is added to the plaintext before encryption. The theory of the decryption is as follows: We apply $\mathbf{f}_k^{-1}$ to both sides of (9.13) to give

$$\mathbf{f}_k^{-1}(\mathbf{c}_i) = \mathbf{f}_k^{-1}(\mathbf{f}_k(\mathbf{p}_i + \mathbf{c}_{i-1})) = \mathbf{p}_i + \mathbf{c}_{i-1}$$

by the definition of $\mathbf{f}_k^{-1}$. Hence

$$\mathbf{p}_i = \mathbf{f}_k^{-1}(\mathbf{c}_i) - \mathbf{c}_{i-1}. \tag{9.14}$$

Finally we note that the first block of plaintext $\mathbf{p}_0$ say may not be properly encrypted, since a simple rule like $\mathbf{c}_i = 0$ for $i < 0$ may be used to fill in the formula at the start. A simple way to solve this problem is to arrange that $\mathbf{p}_0$ is not anything of interest to the eavesdropper. Sensitive information should not be sent until the feedback loop is complete.

9.5 Public key ciphers

9.5.1 *Public key-distribution*

Diffie and Hellman (1976) suggested the possibility of completely new types of cipher. The concepts are theoretically very intriguing, and from a practical point of view they go a long way towards solving the problem of key management. One of these systems is not a cipher in itself, but a way of distributing keys for conventional ciphers 'in public', or at least over an insecure telephone channel. Let us suppose that two people A and B wish to start communication in cipher as soon as possible. They have not previously agreed on a key and there is not time to use a secure courier. Diffie and Hellman offer the following suggestion:

A prime number p and another integer a are selected. Typically p might be of the order of 10^{100}. These integers are assumed to be public knowledge. A chooses a positive integer X (less than p) which he keeps secret. B chooses a number Y in the same way. A computes the value $X' = a^X \bmod p$ and sends it to B, who computes the value $Y' = a^Y \bmod p$ and sends it to A. Then A computes $(Y')^X \bmod p$ and B computes $(X')^Y \bmod p$. Both these values are the same, since

$$(X')^Y \bmod p = (a^X)^Y \bmod p = a^{XY} \bmod p = (a^Y)^X \bmod p = (Y')^X \bmod p.$$

This common value is used as the key. The following points are worth making about this system:

(a) Since the computations are 'modulo p' the eventual key is represented by an integer in the range 0 to $(p-1)$. Hence for sending an n-bit key we need $p > 2^n$.

(b) The function $X' = a^X \bmod p$ is a mapping of the set of integers 0 to $(p-1)$ back into itself. For the highest security the outputs for different values of X should be distinct. A necessary condition is that p should be a prime. Moreover the integer a has to be chosen as 'primitive', which simply means that all its powers 'modulo p' from a^0 to a^{p-1} are distinct. Since in effect we are working in the field $GF(p)$, such a number can always be found. Thus when $p = 11$ we find that $a = 3$ is not primitive since $3^5 = 1 = 3^0 \bmod 11$, and 5 is less than $p - 1$. On the other hand 2 is a primitive element with successive powers 1, 2, 4, 8, 5, 10, 9, 7, 3, and 6 mod 11.

(c) The calculation $a^X \bmod p$ may be performed easily. The exponent X is expressed as a sum of powers of 2, immediately obtained from the representation of X in binary. Then a is repeatedly squared (mod p) to generate a^1, a^2, a^4, a^8, ..., mod p. Then those powers whose exponents add up to X are selected and multiplied (mod p). Thus to compute $31^{55} \bmod 67$ we proceed as follows. First 55 is written as $32 + 16 + 4 + 2 + 1$. Then we compute (mod 67) $31^1 = 31$, $31^2 = 23$, $31^4 = 23^2 = 60$, $31^8 = 60^2 = 49$, $31^{16} = 49^2 = 56$, and $31^{32} = 56^2 = 54$. Then we need $54 \times 56 \times 60 \times 23 \times 31 = 38 \pmod{67}$. (The result of any multiplication may be reduced mod 67 to reduce the size of the numbers involved.) Thus the computing complexity is of order $\log_2 X$, the number of bits in the binary expansion of X. Since X cannot exceed $p - 1$ the complexity is of order $\log_2 p$ or less. Thus values of p of the order of 10^{100} are perfectly feasible.

(d) The eavesdropper must not be able to solve the equation

$$X' \equiv a^X \bmod p, \qquad X', a, \text{ and } p \text{ given.} \tag{9.15}$$

In principle of course he always can solve this equation by trying all possible values for X. But this would involve him in a number of trials of order p, which may well exceed the number of atoms in the known universe. So the question is whether there is a fast algorithm available for this inverse exponentiation. Although a number of algorithms faster than this brute-force method are available in the published literature, none of them is practical for values of p of the order of 10^{100}. The other point to watch is that for certain values of p fast algorithms may be available.

(e) It is quite easy to find primes of the order of 10^{100}. This point is further discussed in Section 9.5.2.

The new feature in this system of public key-distribution is that although there is no theoretical secrecy the eavesdropper cannot find the solution because of the complexity of the computing. The user of such a system is betting that there is no mathematician clever enough to find a fast way of solving (9.15). If such a mathematician exists he may or may not publish his discovery. Thus there is a risk that at any time this system may become totally insecure. Such systems are thus always liable to rumours that someone has broken them, and our eavesdropping friend in his frustration may not be above spreading such rumours in the hope of forcing us to abandon a perfectly good system in favour of something weaker.

The above operations may be regarded as arithmetic operations in $GF(p)$ (Section 7.1.3), with a a primitive element. (It should however be recalled that indices like X and Y are ordinary integers.) The same idea can be used in other fields, provided that a primitive element can be found, not generally an easy task in large fields. In particular the implementations may be easier in a binary extension field $GF(2^m)$. Moreover it may be shown that if $2^m - 1$ is prime then there is no need to search for a primitive element, since every element apart from 0 and 1 is primitive. A reasonable choice for such a field is $GF(2^{127})$ since $2^{127} - 1$ is a prime, one of the Mersenne primes (Pomerance 1982).

So far we have discussed a way of exchanging keys in public. The actual cipher is conventional. In the next section we discuss a cipher-system of the new sort.

9.5.2 *RSA cipher*

The RSA cipher is named after its discoverers, Rivest, Shamir and Adleman (1978). It is based on the idea that whereas primes of the order of 10^{100} are quite easy to find (see below) the factorization of the product of two such primes is infeasible if neither factor is known. The product is of order 10^{200} and thus the usual method of trying every prime up to its square-root (of order 10^{100}) is out of the question. (It is worth pointing out that according to a theorem about prime numbers the fraction of numbers in the vicinity of a large number N that are prime is $1/\ln N$, i.e. about 1 in 230 for $N = 10^{100}$. Thus the number of primes between 10^{99} and 10^{100} is of the order of 10^{97}. There is no way of setting up a table with this number of entries.)

The person B who is going to receive the encrypted messages chooses at random two primes p and q of order 10^{100}. He publishes their product $N = pq$ together with another number E (which satisfies certain weak conditions described below). The person A who wishes to send B an encrypted message first breaks it into blocks which can be represented by positive numbers less than N. Let P be such a block of plaintext. A computes the power

$$C = P^E \bmod N. \tag{9.16}$$

As was pointed out in the last section the complexity of this operation is not excessive. The recipient B computes another power of C,

$$C^D \bmod N = P^{ED} \bmod N. \tag{9.17}$$

The integer D is chosen so that

$$P^{ED-1} \equiv 1 \bmod N. \tag{9.18}$$

In consequence we see that $(\bmod N) P^{ED} = P$, so that the plaintext is recovered. The condition for (9.18) is that $(ED - 1)$ be a multiple of M, where

$$M = (p - 1)(q - 1). \tag{9.19}$$

This follows from a standard result in number theory. Thus we have to make D satisfy the congruence

$$ED \equiv 1 \bmod M. \tag{9.20}$$

Thus by the results of Section 7.6.2 we have to choose E coprime to M, after which an inverse 'modulo M' may be found using the Euclidean algorithm.

The (public) algorithm used by B for setting up this system then has to perform the following tasks:

(a) Two primes p and q have to be chosen at random. The products $N = pq$ and $M = (p - 1)(q - 1)$ are computed.

(b) A number E coprime to M and less than M has to be chosen at random. (Euclid's algorithm may be used to see whether E and M have any common factors.)

(c) A number D is found (again using Euclid's algorithm) satisfying $ED \equiv 1 \bmod M$.

B then publishes N and E as his public key, but he keeps D secret. The primes

p and q are no longer needed. It should be noted that determining M from N is almost as hard as factoring N, since once we know M we immediately obtain $p + q = N - M - 1$. Hence we know the sum and product of p and q, and it is simply a matter of solving a quadratic to find p and q.

It is interesting to note that this cipher provides a solution to the problem of disputes (Section 9.1.1). Suppose B wishes to send A a signed message. For the moment it is assumed that this message need not be encrypted. A needs to be able to show that this message has come from B. He arranges that B encrypts the message using his (B's) secret key D and sends it as

$$P' = P^D \bmod N. \tag{9.21}$$

Anyone, including A, with access to P' can calculate $P'^E \bmod N$ since E and N have been published by B. But this quantity is just P again. A then keeps a copy of P' as well as of P. Anyone can verify that $P'^E \equiv P \bmod N$, but only B could have produced a message P' that can be unscrambled by E. Thus B is publicly shown to be the originator.

Here the message P' is not secret, since anyone can decrypt it. If B wishes to send P secretly but also in its signed condition, then he first computes P' as above and then sends it to A using a cipher system. In particular he may use A's public key to encrypt it, or he may prefer some other way. A then deciphers the truly encrypted message to reobtain P', which he keeps together with the version P obtained by using B's public key.

There are several points worth making about this system:

(a) Finding prime numbers is not too difficult (Pomerance 1982). The quickest way is to submit a candidate number to a succession of tests. Each time it passes such a test the chances of its not being prime are halved. Thus after it has passed say a hundred tests the probability of its not being prime is of order 2^{-100}. This should be safe enough for most purposes, and further testing can be used if required. The other method is very slow, but does guarantee with mathematical certainty that a number is prime.

(b) The modular computations need hardware considerably more complicated than that used by other ciphers, since the arithmetic is not binary. Thus the system tends to be rather slow.

(c) There is still the hazard that the system may be broken by a brilliant mathematician. Conventional ciphers use fairly messy 'scrambling' algorithms, with no very apparent mathematical structure. Thus one might feel that they are less liable to succumb to mathematical ingenuity than the mathematically highly structured public key ciphers.

For further reading: Kahn (1967) gives a history of cryptography. Beker and Piper (1982) give an overview at a moderately technical level. Meyer and Matyas (1982) is a text at a more advanced level, with special emphasis on the Data Encryption Standard. Diffie and Hellman (1979) and Sloane (1981a) have written articles that are easy to read and yet are very instructive.

Exercises

9.1 The polynomial $Q(z)$ over a given finite field is not a multiple of z. Show that it divides (is a factor of) $1 - z^n$ for some positive n, as follows: (a) Show that without loss of generality it may be assumed that $Q(0) = 1$. (b) Show that a shift-register as in Section 9.3.3 may be set up to divide 1 by $Q(z)$ with the initial state (i.e. contents) of the register set to $1 = (1000 \ldots)$. (c) Show that if in the sequence of successive states $\mathbf{u}_1(=1), \mathbf{u}_2, \mathbf{u}_3, \ldots$ the state $\mathbf{1}$ appears again, then the result is quickly proved. (d) Show that there must be a state that is repeated in the sequence. (e) Let l be the least integer such that $\mathbf{u}_l$ has appeared previously, at location k. Show that if $\mathbf{1}$ is never repeated then $k > 1$. (f) Show that $\mathbf{w} = \mathbf{u}_{l-1} - \mathbf{u}_{k-1}$ is non-zero and that its successor is zero. (h) Use contradiction to show that $\mathbf{1}$ must be repeated.

Show also that if n is the least positive integer for which $Q(z)$ divides $1 - z^n$ then the dividend is the polynomial given by the first period of the long division of $Q(z)$ into 1, and that n is the length of this period.

9.2 Show that if $N = 2^n - 1$ is divided by $M = 2^m - 1$ (with m and n positive integers), then the remainder is $R = 2^r - 1$, where r is the remainder when n is divided by m. Hence show that the h.c.f. of M and N is $2^h - 1$ where h is the h.c.f. of m and n. Hence or otherwise show that N is not prime if n is not prime.

9.3 Show that the permutation of three-bit strings into three-bit strings given by $0 \to 6, 1 \to 1, 2 \to 5, 3 \to 2, 4 \to 7, 5 \to 0, 6 \to 4$, and $7 \to 3$ is affine, and express the transform in the form $\mathbf{c} = \mathbf{Ap} + \mathbf{b}$ where $\mathbf{A}$ is a 3×3 matrix and $\mathbf{b}$ a binary vector. (Here the strings are represented by their octal equivalents; thus 6 stands for 110.)

9.4 Show that the number of invertible binary 5×5 matrices is

$$(2^5 - 1)(2^5 - 2)(2^5 - 2^2)(2^5 - 2^3)(2^5 - 2^4) = 9\,999\,360.$$

(Hint: The first column can be chosen any way except for all-zeros. The second column must not be a multiple of the first, the third must not be a linear combination of the first two, and so on.) Hence show that the fraction of all binary 5×5 matrices which are invertible is 29.8 per cent.

9.5 Show that as $n \to \infty$ the fraction of all binary $n \times n$ matrices that are invertible tends to $(1 - 1/2)(1 - 1/4)(1 - 1/8) \ldots = 0.28$.

9.6 Given m randomly chosen n-vectors (column vectors) with $m \geq n$ show that the chances of finding a linearly independent subset of them of size n is $(1 - 2^{-m})(1 - 2^{1-m}) \ldots (1 - 2^{n-1-m})$. (Hint: Form the columns into an $n \times m$ matrix. Use the theorem that the maximum number of linearly independent columns is the same as the maximum number of linearly independent rows.) What is the answer for $m = 6, n = 4$?

9.7 Show that the total number of invertible n-bit to n-bit transforms is $(2^n)!$

9.8 Show that the total number of invertible affine transforms of n bits into n bits is

$$2^n(2^n - 1)(2^n - 2)(2^n - 2^2)(2^n - 2^3) \ldots (2^n - 2^{n-1}).$$

9.9 Hence show that the fraction of invertible four-bit to four-bit transforms that are affine is 1.54×10^{-8}, and three-bit to three-bit is 0.0333.

Appendix A. Exact error probability for biorthogonal codes

An exact formula for the error probability of a $(2^{k-1}, k)$ Reed-Muller code is derived. This code is a biorthognal code like the one illustrated in Fig. 5.2, but with 2^k signal-points in $M = 2^{k-1}$ dimensions. The amplitude c is the square-root of the total energy kE_b of each word, given by multiplying the energy per message-bit by the number of such bits in each block. Let us assume without loss of generality that the signal S_1 is sent. A correct decoding occurs if (a) the measured amplitude z of the first coordinate of the received signal R is positive, and (b) the amplitudes of all the other $M - 1$ coordinates of R are less than z in magnitude. The probability of the amplitude of the first coordinate lying in the range z to $z + dz$ is $(\pi\eta)^{-1/2} \exp\{-(z-c)^2/\eta\}\, dz$. The probability that any other coordinate has a magnitude less than z is

$$\int_{-z}^{z} (\pi\eta)^{-1/2} \exp(-t^2/\eta)\, dt = \operatorname{erf}(z/\sqrt{\eta}).$$

Since z must be positive the probability of correct decoding is

$$\int_0^\infty \{\operatorname{erf}(z/\sqrt{\eta})\}^{M-1} (\pi\eta)^{-1/2} \exp\{-(z-c)^2/\eta\}\, dz$$

with;

$$c = \sqrt{(kE_b)}, \qquad M = 2^{k-1}. \tag{A.1}$$

It is convenient to substitute z for $z/\sqrt{\eta}$. In terms of the error probability P_W the probability of correct decoding is $1 - P_W$. Hence we find

$$1 - P_W = \int_0^\infty \{\operatorname{erf}(z)\}^{M-1} F'(z - c/\sqrt{\eta})\, dz$$

$$= 1 - \int_0^\infty [\{\operatorname{erf}(z)\}^{M-1}]'\, F\{z - \sqrt{(k\gamma_b)}\}\, dz \tag{A.2}$$

after integrating by parts and introducing $\gamma_b = E_b/\eta$ (eqn (6.13)). Here the prime denotes differentiation with respect to z and $F(z)$ is the Gaussian cumulative distribution function

$$F(z) = \pi^{-1/2} \int_{-\infty}^z \exp(-t^2)\, dt = \tfrac{1}{2}\{1 + \operatorname{erf}(z)\},$$

so that $F'(z) = \pi^{-1/2} \exp(-z^2)$. The integral on the right-hand side of (A.2) thus gives P_W. Its numerical evaluation is straightforward for k less than 10, if a little care is exercised.

Appendix B. Justification of the Euclidean algorithm

Some preliminaries

(a) As in Section 7.6.2 we use the cross-product theorem to prove

$$P_i Q_{i+1} - P_{i+1} Q_i = (-1)^i G, \qquad i = 0, 1, 2, \ldots \tag{B.1}$$

and

$$Q_i R_{i+1} - Q_{i+1} R_i = (-1)^{i+1}. \tag{B.2}$$

(b) Table 7.8 shows that P_i is the linear combination

$$P_i = G R_i + F Q_i. \tag{B.3}$$

(c) We need the theorem that, if F is relatively prime to G and is a factor of KG, then F is a factor of K. To show this we multiply both sides of (7.25) by K, and put $KG = AF$ for some A to give $F(AM + KL) = K$, so that F is a factor of K.

(d) From this follows the *unique factorization theorem*, that a polynomial can be factored into irreducible factors in essentially only one way. For suppose that a polynomial can be factored in two ways and that F is an irreducible factor in one factorization relatively prime to all the irreducible factors $G_1, G_2, \ldots, G_r$ in another. Then since F is relatively prime to G_r it must be a factor of $G_1 G_2 \ldots G_{r-1}$, and iterating this argument shows that it must be a factor of G_1, which is a contradiction.

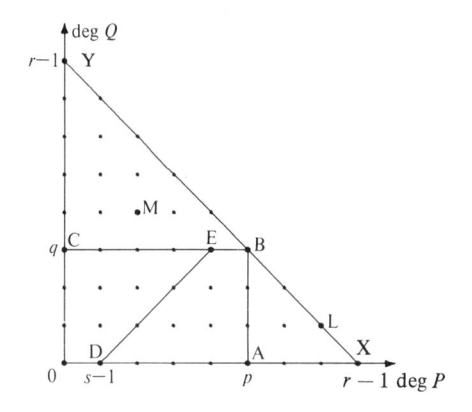

FIG. B.1 Operation of Euclidean algorithm.

Main part of justification

(a) We consider polynomial pair solutions (P, Q) of the key equation

$$P \equiv QF \bmod G \tag{B.4}$$

regardless of their degrees for the moment. Here F is $(S \bmod G)$ or $RS \bmod G$ as described in Section 7.7.3 or Section 7.7.4. We picture a given solution (P, Q) as 'lying' at the point (deg P, deg Q) in a plot like Fig. B.1. Evidently these solutions are not unique since P and Q may both be multiplied by an arbitrary polynomial.

(b) We start by considering the solutions (P_i, Q_i) produced by the Euclidean algorithm (Section 7.7.3). Plotted in the figure they move diagonally upward and to the left as i is increased. The first point (P_1, Q_1) is usually at X since F is usually of degree $r - 1$. (It may be less.) Moreover deg P_i is strictly decreasing with i by (7.19) and deg Q_i is strictly increasing by (7.21). Thus by (B.1) we have

$$\deg P_i + \deg Q_{i+1} = \deg G = r \tag{B.5}$$

and hence

$$\deg P_i + \deg Q_i < r.$$

In particular these solutions always lie in the triangle OXY (the boundary of this and other shapes counting as 'in'). We now choose non-negative integers p and q satisfying $p + q = r - 1$ and we consider the 'pq-rectangle' $OABC$ of dimensions p by q. Then it is easy to show that the algorithm produces one and only one solution in any such rectangle. For suppose that the ith solution is at a point like L, to the right of the rectangle, and the $(i + 1)$th is at a point like M above the rectangle. Then we find deg $P_i \geqslant p + 1$ and deg $Q_{i+1} \geqslant q + 1$. Thus deg P_i + deg $Q_{i+1} \geqslant p + q + 2 = r + 1$, which contradicts (B.5). On the other hand, if L and M are both in the rectangle, then we find dep P_i + deg $Q_{i+1} < r$, again contradicting (B.5).

(c) The polynomials P and Q of the true solution given by (7.38) and (7.39) are relatively prime. For the irreducible factors of Q are of the form $(z - x_k)$, where x_k corresponds to an error, so that y_k is not zero. But we see that for given k the factor $(z - x_k)$ appears in every term of $P(z)$ but one, and hence $P(z)$ is not a multiple of $(z - x_k)$.

(d) The number of erasures s is of course known. We assume that the number of errors t satisfies $s + 2t \leqslant r$. We now set q in Fig. B.1 equal to the integer part of $\frac{1}{2}(r - s)$ and set $p = (r - 1) - q$. Then the true solution must lie in the trapezium $ODEC$ and hence in that particular pq-rectangle $OABC$. For we immediately find $t \leqslant q$, so that the true solution (P, Q) lies on or below CB. Moreover if we replace the inequality in (7.42) by deg $P = s + t - 1$, then (P, Q) must lie on DE, and by this inequality it can lie to the left as well.

(e) There are no other solutions of the key equation (B.4) in this rectangle except those of the form (KP, KQ). For suppose there is another solution (U, V) in this rectangle, that is with deg $U \leq p$, deg $V \leq q$. Then if we multiply both sides of $P \equiv QF$ mod G by V and both sides of $U \equiv VF$ mod G by Q and then subtract we obtain $(PV - QU) \equiv 0$ mod G. But the degree of the left-hand side is less than r, the degree of G, and so we obtain the polynomial equality $PV - QU = 0$, or $PV = QU$. So P is a factor of QU and since it is relatively prime to Q it must be a factor of U. Hence we may set $U = KP$, so that $PV = QKP$, from which by cancellation we find $V = KQ$.

(f) Thus the Euclidean algorithm must give a solution of the form (P, Q) in this rectangle, or a multiple (KP, KQ). The latter possibility does not happen. For suppose $P_i = KP$, $Q_i = KQ$, where (P, Q) is the true solution. From (B.3) we find $KP = GR_i + FKQ$. Moreover, since by (B.4) $P - FQ$ is a multiple of G, say AG, we may substitute $P = FQ + AG$ to give $KAG = GR_i$, from which we obtain $KA = R_i$ by cancellation. Thus K is a factor of R_i. It is also a factor of Q_i since $Q_i = KQ$, and so it is a common factor of Q_i and R_i. But then it is a factor of the left-hand side of (B.2), which is a scalar. So K is also a scalar.

(g) The true solution which must be obtained by the algorithm then obeys deg $P_i < s +$ deg Q_i, but the previous solution (P_{i-1}, Q_{i-1}) must lie to the right of AB in Fig. B.1 and so does not obey this condition. Of course, if the Euclidean solution in the rectangle is to the right of DE, then this means a decoding failure due to too many errors.

(h) Finally we consider the need for (7.41). If this inequality is violated, then it is possible to have a solution outside OXY, and thus inaccessible to the Euclidean algorithm.

A note on the Berlekamp algorithm: In the Euclidean algorithm for solving the key equation the whole of the syndrome polynomial S (or F) is included in the calculation from the start. The successive solutions (P_i, Q_i) in Fig. B.1 start at X and appear successively upwards and to the left. In the Berlekamp algorithm the higher terms from $S(z)$ are taken into the calculation only as needed. The solutions in Fig. B.1 start at O and appear successively upwards and to the right. The algorithm is easier than the Euclidean algorithm to implement, but it is not so easy to prove that it works.

References

Abramowitz, M. and Stegun, I. A. (eds.) (1965). *Handbook of mathematical functions.* Dover.
Allen, C. W. (1973). *Astrophysical quantities* 3rd edn. Athlone.
American Institute of Physics Handbook, 3rd edn. (1972). McGraw-Hill.
Beker, H. and Piper F. (1982). *Cipher systems.* Northwood.
Berlekamp (1980). The technology of error-correcting codes, *Proc. IEEE* **68**, 564-93.
Bhargava, V. K. (1983). Forward error correcting schemes for digital communications. *IEEE Commun. Mag.* **23**, No. 1, 11-19.
Bhargava, V. K., Haccoun, D., Matyas, R., and Nuspl, P. (1981). *Digital communications by satellite: modulation, multiple access and coding.* Wiley.
Blahut, R. E. (1983). *The theory and practice of error-control codes.* Addison-Wesley.
Blake, I. F. and Mullin, R. C. (1975). *The mathematical theory of coding.* Academic.
Bracewell, R. N. (1978). *The Fourier transform and its applications* 2nd edn. McGraw-Hill Kogakusha.
Burden, R. L., Faires, J. D., and Reynolds, A. C. (1981). *Numerical analysis* 2nd edn. Prindle, Weber and Schmidt.
Bylanski, P. and Ingram, D. G. W. (1980). *Digital transmission systems* revised edn. IEE.
Callahan, M. B. (1981). Submarine communications. *IEEE Commun. Mag.* **19**, No. 6, 16-25.
Carasso, M. G., Peek, J. B. H., and Sinjou, J. P. (1982). The Compact Disc audio system. *Philips Tech. Rev.* **40**, 151-6.
Chase, D. (1972). A class of algorithms for decoding block codes with channel measurement information. *IEEE Trans. Inf. Theory* **IT-18**, 170-82.
Chung, K. L. (1974). *Elementary probability theory with stochastic processes.* Springer.
Clark, G. C. and Cain, J. B. (1981). *Error-correcting codes for digital communications.* Plenum.
Coates, R. C. (1983). *Modern communication systems,* 2nd edn. Macmillan.
Cook, C. E. and Marsh, H. S. (1983). An introduction to spread-spectrum. *IEEE Commun. Mag.* **21**, No. 2, 8-16.
Davisson, L. D. and Gray, R. M. (eds). (1976). *Data compression.* Dowden, Hutchinson and Ross.
Delsarte, P. and Piret, P. (1982). Algebraic construction of Shannon codes for regular channels. *IEEE Trans. Inf. Theory* **IT-28**, 593-9.
Diffie, W. and Hellman, M. E. (1976). New directions in cryptography. *IEEE Trans. Inf. Theory* **IT-22**, 644-54.
Diffie, W. and Hellman, M. E. (1979). Privacy and authentication: an introduction to cryptography. *Prof. IEEE* **67**, 397-427.
Dixon, R. C. (1976). *Spread spectrum systems.* Wiley.
Conan Doyle, Sir Arthur. (1981). The adventure of the dancing men. pp. 511-26 in *The Penguin complete Sherlock Holmes.* Penguin.
Edelson, R. E., Madson, B. D., Davis, E. K., and Garrison, G. W. (1978).

Voyager telecommunications: the broadcast from Jupiter. *Science* **204**, 913-21.

Farrell, P. G. (1979). Array codes. Chapter 4 in *Algebraic coding theory and applications* (ed. G. Largo). Springer-Verlag.

Feher, K. (1981). *Digital communications: microwave applications.* Prentice-Hall.

Feher, K. (1983). *Digital communications: satellite/earth station engineering.* Prentice-Hall.

Feistel, H. (1973). Cryptography and computer privacy. *Sci. Am.* **228**, 15-23.

Forney, G. D. Jr. (1966). *Concatenated codes.* MIT.

Forney, G. D. Jr. (1981). Burst correcting codes for the classic bursty channel. *IEEE Trans. Commun.* COM-19, 772-81.

Gagliardi, R. M. (1978). *Introduction to communications engineering.* Wiley.

Gallager, R. G. (1968). *Information theory and reliable communication.* Wiley.

Garrett, I. (1983). Towards the fundamental limits of optical-fiber communications. *J. Lightwave Technol.* **1**, 131-8.

Gersho, A. (1977). Quantization. *IEEE Commun. Mag.* **17**, No. 5, 16-28.

Gersho, A. and Cuperman, V. (1983). Vector quantization—a pattern matching technique for speech coding. *IEEE Commun. Mag.* **21**, No. 9, 15-21.

Gradshteyn, I. S. and Ryzhik, I. M. (1965). *Tables of integrals, series and products* 4th edn. Academic.

Graham, J. (1983). *The Penguin dictionary of telecommunications*, Penguin.

Hamming, R. W. (1980). *Coding and information theory.* Prentice-Hall.

Harmuth, H. F. (1981). *Non-sinusoidal waves for radar and radio communication.* Academic.

Haykin, S. (1978). *Communication systems.* Wiley.

Hoel, P. G., Port, S. B., and Stone, C. J. (1971). *Introduction to probability theory.* Houghton-Mifflin.

Hoeve, H., Timmermans, J., and Vries, L. B. (1982). Error correction and concealment in the Compact Disc system. *Philips Tech. Rev.* **40**, 166-72.

Holmes, J. K. (1982). *Coherent spread spectrum systems.* Wiley.

Howes, M. J. and Morgan, D. V. (eds.) (1980). *Optical fibre communications.* Wiley.

IEEE Trans. Commun. COM-30, No. 4, part 1, (April 1982). Special issue on bit-rate reduction.

IEEE Trans. Commun. COM-30, No. 5, part 1, (May 1982). Special issue on spread-spectrum communications.

ITT (1975). *Reference data for radio engineers.* H. W. Sams.

Justesen, J. (1976). On the complexity of decoding Reed-Solomon codes. *IEEE Trans. Inf. Theory* IT-22, 237-8.

Kahn, D. (1967). *The codebreakers: the story of secret writing.* Macmillan.

Kaye, G. W. C. and Laby, T. H. (1972). *Tables of physical and chemical constants* 14th edn. Longman.

Keiser, G. (1983). *Optical fiber communications.* McGraw-Hill.

Lathi, B. P. (1983). *Modern digital and analog communication systems.* Holt-Saunders.

Lesh, J. R. (1982). Optical communications research program to demonstrate 2.5 bits/detected photon. *IEEE Commun. Mag.* **20**, No. 6, 35-7.

Lighthill, M. J. (1959). *Introduction to Fourier analysis and generalized functions.* Cambridge.

Lin, S. and Costello, D. J. (1983). *Error control coding: fundamentals and applications.* Prentice-Hall.

REFERENCES

Lucky, R. W., Salz, J., and Weldon, E. J. (1968). *Principles of data communication.* McGraw-Hill.
McEliece, R. J. (1977). *The theory of information and coding: a mathematical framework for communication.* Addison-Wesley.
MacWilliams, F. J. and Sloane, N. J. A. (1977). *The theory of error-correcting codes* Parts I and II. North-Holland.
Martin, J. (1976). *Telecommunications and the computer.* Prentice-Hall.
Martin, J. (1977). *The future of telecommunications.* Prentice-Hall.
Martin, J. (1978). *Communication satellite systems.* Prentice-Hall.
Massey, J. L. (1969). Shift-register synthesis and BCH decoding. *IEEE Trans. Inf. Theory* **IT-15**, 122-7.
Meyer, C. H. and Matyas, S. M. (1982). *Cryptography.* Wiley.
Midwinter, J. E. (1979). *Optical fibers for transmission.* Wiley.
Miller, G. M. (1983). *Modern electronic communication.* Prentice-Hall.
Mimms, F. W. III (1982). *A practical introduction to lightwave communications.* H. W. Sams.
National Bureau of Standards (1977). Data encryption standard. Federal Information Processing Standard (FIPS) Publication No. 46.
Oppenheim, A. V. and Schafer, R. W. (1975). *Digital signal processing.* Prentice-Hall.
Parzen, E. (1980). *Modern probability theory and its applications.* Wiley.
Pasupathy, S. (1977). Correlative coding: a bandwidth-efficient signaling scheme. *IEEE Commun. Mag.* **15**, No. 4, 4-11.
Pasupathy, S. (1979). Minimum shift keying: a spectrally efficient modulation. *IEEE Commun. Mag.* **19**, No. 4, 14-22.
Peebles, P. Z. (1976). *Communication system principles.* Addison Wesley.
Personick, S. D. (1981). *Optical fiber transmission systems.* Plenum.
Philips Tech. Rev. Vol. 40, (1982). No. 6, 151-80. Issue on Compact Disc digital audio system.
Pierce, J. R. and Posner, E. C. (1980). *Introduction to communication science and systems.* Plenum.
Pomerance, C. (1982). The search for prime numbers. *Sci. Am.* **247**, 122-30.
Proakis, J. G. (1983). *Digital communications.* McGraw-Hill.
Qureshi, S. (1982). Adaptive equalization. *IEEE Commun. Mag.* **20**, No. 2, 9-16.
Reif, F. (1965). *Statistical physics.* Berkeley Physics Course Vol. 5. McGraw-Hill.
Rivest, R. L., Shamir, A., and Adleman, L. (1978). A method for obtaining signatures and public-key cryptosystems. *Commun. ACM* **21**, 120-6.
Roddy, D. and Coolen, J. (1981). *Electronic communications,* 2nd edn. Reston.
Roden, M. S. (1979). *Analog and digital communication systems.* Prentice-Hall.
Roden, M. S. (1982). *Digital and data communication systems.* Prentice-Hall.
Rosie, A. N. (1973). *Information and communication theory,* 2nd edn. Van Nostrand Rheinhold.
Sarwate, D. V. (1977). On the complexity of decoding Goppa codes. *IEEE Trans. Inf. Theory* **IT-23**, 515-16.
Schrader, R. L. (1980). *Electronic communications* 4th edn. McGraw-Hill.
Schwartz, M. (1980). *Information transmission, modulation and noise* 3rd edn. McGraw-Hill.
Schwartz, M., Bennett, W. R. and Stein, S. (1966). *Communication systems and techniques.* McGraw-Hill.
Shamir, A. (1979). How to share a secret. *Commun. ACM* **22**, 612-13.
Shanmugam, K. S. (1979). *Digital and analog communication systems.* Wiley.

Shannon, C. E. (1948). A mathematical theory of communication. *Bell Syst. Tech. J.* **27**, 379–423 and 623–56.
Sklar, B. (1983a). Tutorial on digital communications part 1. *IEEE Commun. Mag.* **21**, No. 5, 4–17.
Sklar, B. (1983b). Tutorial on digital communications part 2. *IEEE Commun. Mag.* **21**, No. 7, 6–21.
Sloane, N. J. A. (1981a). Error-correcting codes and cryptography. In *The Mathematical gardner* (D. Klarner, ed.). Wadsworth.
Sloane, N. J. A. (1981b). Tables of sphere packings and spherical codes. *IEEE Trans. Inf. Theory* **IT-27**, 327–38.
Sloane, N. J. A. (1984). The packing of spheres. *Sci. Am.* **250**, 92–101.
Spiegel, M. R. (1964). *Theory and problems of complex variables.* Schaum's Outline Series, McGraw-Hill.
Spilker, J. J. (1977). *Digital communication by satellite.* Prentice-Hall.
Stafford, R.H. (1980). *Digital television: bandwidth reduction and communication aspects.* Wiley-Interscience.
Stanley, W. D. (1982). *Electronic communication systems.* Reston.
Stark, H. and Tuteur, F. B. (1979). *Modern electrical communications.* Prentice-Hall.
Stremler, F. G. (1982). *Introduction to communication systems* 2nd edn. Addison Wesley.
Sugiyama, Y., Kasahara, M., Hirawawa, S., and Namekawa, T. (1976). An erasures- and errors-decoding algorithm for Goppa codes. *IEEE Trans. Inf. Theory* **IT-22**, 238–41.
Tomlinson, M. (1983). Satellite TV system has digital-analogue phase modulation. *Wireless World* **89**, 28–9.
Wichmann, E. I. (1972). *Quantum physics.* Berkeley Physics Course Vol. 4. McGraw-Hill.
Wiggert, D. (1978). *Error control coding and applications.* Artech House.
Wolf, J. K. (1978). Efficient maximum likelihood decoding of linear block codes using a trellis. *IEEE Trans. Inf. Theory* **IT-24**, 76–80.
Wozencraft, J. M. and Jacobs, I. M. (1965). *Principles of communication engineering.* Wiley.
Ziemer, R. E. and Tranter, W. H. (1976). *Principles of communications.* Houghton Mifflin.

Index

See also glossary of function names and symbols on p. xv.

affine transform   216–17, 218, 226, 227
AM (amplitude modulation)   6, 60
analogue signalling   2, 3
   v. digital signalling   2–3, 55
analogue-to-digital conversion   49–51, 107
antenna gain   8–9
ARQ (automatic repeat request)   125, 145–6
array code   123–4, 125, 128, 148
ASCII code   41, 42, 206
ASK (amplitude-shift keying)   64, 65
authentication   203, 215
autocorrelation function
   in Fourier theory   19
   in probability theory   78, 79
   of noise   80, 81, 120
AWGN (additive white Gaussian noise) channel   86, 88, 195
   codes on,   126, 127, 140–3, 145
   reliable communication on   87–89, 111–14

band-limited signal   14, 52
bandpass filter   6
bandwidth   6, 10
bandwidth equivalent   95, 200
baseband signalling   43–52
basis set   94, 97, 119, 199
BCH (Bose-Ray-Chaudhuri-Hocquenghem) code   159–60, 166, 169, 177
Berlekamp's algorithm   178, 231
BFSK (binary frequency-shift keying)   65
biorthogonal signalling, see orthogonal signalling
bit (unit of information)   40, 41
bit-error rate   87, 143
black-body radiation   10, 82; see also noise, thermal
block code   123, 133–40, 147
Boltzmann's constant   10, 82
BPSK (binary phase-shift keying)   64
BSC (binary symmetric channel)   122–3, 136, 149; see also decoding, hard-decision
burst channel   90, 119, 127
burst-error-correcting code   91, 127–8

camouflaging a transmission   115
cancellation law (algebraic)   154, 161, 163, 173
carrier   6
carrier synchronization   60, 63, 64, 65
carrier-based signalling   60–6
carrier-phase tracking   65, 90; see also carrier synchronization
causality condition   21, 25
CCITT   165
CDM (code-division mutiplexing)   116
central limit theorem   76, 92
channel capacity   4, 113–15
   quantum limit on   197–9
channel, memoryless   123, 127, 191
check-sum   148, 158, 168–9, 215; see also parity check
Chernoff bound   74–5, 181
cipher   203
   bigram   204–5
   block   205, 216–20
   DES   219–20
   Feistel   219
   monoalphabetic   204, 205
   polyalphabetic   205
   public-key   204, 221–5
   RSA (Rivest–Shamir–Adleman)   223–5
   running-key   205, 209–10
   stream   205, 211–16
cipher block-chaining   221
cipher feedback   220–1
ciphers, avalanche effect in   218
closure property (algebraic)   152, 153, 156, 157
code, (n, k) 123, 167–8
codeword   104
coding gain   141
coding, source, see data-compression
coherent optical reception   192–5, 196, 200
combinatorial factor   70–1
communications, optical   8, 184
Compact Disc, see digital audio system
compander   50
concatenated code   142, 143, 178–82, 191
conditional probability   71

INDEX

congruence 162, 163; see also 'mod' symbol
constraint length 133
contour integration 15
convergence factor 28; see also windowing
convolution 18, 22, 25, 47, 53
convolutional code 132-3, 181
Cooley-Tukey algorithm 100
correlation function 19
covariance 72, 73
CPSK (continuous phase-shift keying) 131, 132
CRC (cyclic redundancy check) 164
cross-product theorem 170
cyclic code 165

data compression 41-3
   irreversible 42, 50, 51
dB (decibel) 9
DBS (direct broadcast from satellite) 7
decision procedure 47
decoder 47
decoding
   Chase 146-7
   complexity of, 89, 114, 180-1
   error-trapping 140
   hard-decision 122, 134, 143, 144, 145, 146
   majority-logic 139
   minimum-Euclidean-distance 105
   minimum-Hamming-distance 135, 136
   soft-decision 126, 128, 142, 146, 178
   syndrome 137-8, 159-60, 174-5
   Viterbi 130, 147-8
decryption (deciphering) 203
degree kelvin 10
Delsarte-Piret code 181
delta comb 34-6
delta function 19-20
delta symbol (Kronecker) 79, 96
demodulator 47
   synchronous 63
DES (Data Encryption Standard) 219-20
DFT (discrete Fourier transform) 38
digital audio system 3, 128
digital signalling 2-3, 39, 43-4, 50, 55
disputes, problem of 225
DMC (discrete memoryless channel) 191
dot product 100
DPSK (differential phase-shift keying) 65, 90
duality (in Fourier theory) 16, 18, 54
duobinary code 58-9, 67, 128-9

eavesdropper 203
encoder 46

encryption (enciphering) 203
energy of signal 98
energy theorem 19
energy-sphere 104, 106, 107
ensemble 69, 78
envelope demodulation 60
equalizer 56, 186, 195
   adaptive 56-8
erasure 167, 176-7, 191
erf (error function) 228
erfc (error-function complement) 76-7
ERP (effective radiated power) 9
error-detection 123, 125, 145-6, 164-5, 219
Euclidean algorithm 169-72, 174, 176, 177, 178, 229-31
Euclidean distance 96, 105, 126
event in probability theory 69-70
   independent 70
   exclusive 70
expansion with a basis 97-8
experiment 69-70
exponential dependence 28
extended code 140, 148

fading channel 89-91
false alarm 189
'fast' algorithm 100, 180
FDM (frequency division multiplexing) 60
feedback shift-register 164-5, 211-12
fibre
   monomode 185, 186
   multimode 186
fibre-optic transmission 184, 185-6, 188, 189, 201
field (algebraic) 153-5, 223
   extension 155-7, 163, 172
filter 20-6, 47, 48
   bandpass 6, 26, 192, 197
   impulse-response function of 21-2, 23, 25
   low-pass 24-6, 49
   matched 101-2
   sliding-average 24, 49, 55, 86; see also integrate-and-dump process
   transfer function of 22-3
   transversal 56, 57
Fire code 127
Fourier coefficient 27
Fourier integral 13
Fourier series 26-9
Fourier transform 13, 14, 17
   inverse 13
frequency 6
frequency bands 6-8, 184
frequency spectrum 14

238   INDEX

FSK (frequency-shift keying)   65, 90

gain in coherent optical reception   193
Gaussian distribution   76-7
Gaussian function   76
Gaussian noise   81, 85-6
    as limiting information rate   114
    error-probability with,   87
    in polar signalling   86-9
    in signal-space   103
generator matrix   134
geostationary satellite   11
GF (Galois field), see field (algebraic)
Golay code   140, 149, 166
'good' function   15
Goppa code   178
Gray code   45, 46

Hadamard function   99
    as Reed-Muller code   124-5
Hadamard matrix   99
Hamming code   138, 149, 157-9, 166
    error probability with   144, 145
Hamming distance   133, 140
Hamming weight   133
h.c.f. (highest common factor)   169, 170
Heisenberg uncertainty principle   196
HF (high frequency) band   7
Huffman code   41

induction, mathematical   44, 67, 100
information   40-1
information rate   40-1
information theory   43
integrate-and-dump process   44, 55, 96, 102; see also filter, sliding-average
interference, see noise
interleaved code   128
inverse 'mod G'   171-2
inverse of field element   153
ionosphere   7
ISI (intersymbol interference)   39, 56

jamming   115

key (cipher)   203
    base   216
    message   216
    see also cipher, public-key
key-distribution, public   221-3
key-stream   205, 211
key equation   175
known-plaintext attack   210-11
Kronecker delta symbol   79, 96

laser   184, 189, 195

lattice code   106-7
laws of algebra   152-3
Leech lattice   107
light, visible   8, 10, 184
limits on information rates   113, 114, 197-9, 200-1
linear phase   38
looped wave-guide   32-4, 78, 82, 198
loudspeaker   12

magnitude, stellar   9-10
mapping   206-7
Mersenne prime   223
message block   123
metric   129
microwave   8
Miller code   67
MLD (maximum-likelihood decoding)   104, 135
'mod' symbol   123, 152, 162; see also congruence
modular arithmetic   152; see also congruence, 'mod' symbol
modulation   6
    angle   65, 118, 130-2
    delta   51
    quadrature   63
    see also AM, ASK, BFSK, BPSK, CPSK, DPSK, FSK, MPSK, MSK, OOK, PAM, PCM, PPM, PRK, PSK, QAM, QASK, QPSK
modulator   46
Morse code   41, 42, 46
MPSK (multiple phase-shift keying)   64, 65
MSK (minimum-shift keying)   130-1, 132
M-sphere (hypersphere)   104, 112
multi-h phase code   130-2
multivariate distribution   77
mutually Gaussian distribution   77

nat (unit of information)   40
noise (interference)   11, 44
    filtered   80-2
    quantization   50
    shot   82-5
    thermal   10, 82, 85, 185, 196-7
    white   80, 102-3
    see also Gaussian noise
noise energy   81, 82, 197
noise spectral density   79, 81
noise-temperature   10, 82, 187, 188, 195
normal mode   33, 82, 199, 200
normalization   77, 96, 155, 197
Nyquist property   36, 54
Nyquist sampling   47-8

INDEX

one-time pad (or tape)   207-8, 209, 210
OOK (on–off keying)   43, 44, 64, 190
optic fibres   185-6
optical heterodyne reception   192-5
orthogonal functions   96
orthogonal signalling   107-11, 120, 142, 150
orthonormal set   97
outcome (of experiment)   69
outer code   180, 181

PAM (pulse-amplitude modulation)   50, 55
parity check   123, 124; *see also* checksum
parity-check matrix   137, 139, 140, 158
Parseval's theorem   19, 30
PCM (pulse-code modulation)   50, 55
performance (of certain codes)   142, 145
period   6, 45 n.
periodic function   26, 30, 34
periodically repeated function   32
phasor   62
photodiode   187, 188
    avalanche   188, 189
photoelectric effect   187
photomultiplier   188
photon   185, 187, 196, 199-200
    detection of   185, 187-90, 196, 197, 199
photon rate   191
pipelining   218, 221
plaintext   203
Planck's constant   187
Poisson process   82
polynomial   160-1, 213
    derivative of   183
    division algorithm for   161-2, 164, 165, 212-13
    error-locator   175, 177
    factor of   161
    generator   163
    irreducible   161
    multiple of   161
    quotient   161
    remainder   161
positive-frequency representation   61
power   8-9
power flux   8
power spectral density   79
power theorem   30
power-limited signalling   117
PPM (pulse-position modulation) signalling   88, 92, 190-2
    as orthogonal signalling   107
primitive element   155, 222, 223

PRK (phase-reversal keying)   64
probability density   73-4
    Gaussian   76
    joint   73
PSK (phase-shift keying)   64

QAM (quadrature-amplitude modulation)   64
QASK (quadrature amplitude-shift keying)   64
QPSK (quadrature phase-shift keying)   64
quantization   50
quantum, *see* photon
quantum theory   8, 32, 178, 192, 194, 195-200
quasi-rectangular pulse   53, 141, 199

raised-cosine function   15
random variable(s)
    continuous   73
    covariance of   72
    discrete   72, 74
    i.i.d.   74
    independent   74, 77
    mean of   72
    uncorrelated   73, 74, 76, 77
    variance of   72
Rayleigh fading   89-91
Rayleigh scattering   186
reality condition   16, 23, 30
recipient   203
redundancy
    in codeword   174
    in data   42, 59, 167, 210
Reed-Muller code (first order)   124-5, 127, 135
    as orthogonal system   141
    related to cyclic codes   166
    use in concatenated codes   179, 180
    word-error probabilities   141-3, 144, 145, 146, 228
Reed-Solomon code   127, 128, 167-9, 174, 175, 191-2
    in concatenated codes   178-81
    in digital audio system   128
reliability
    arbitrarily high   4, 87-9, 110, 113, 180-2
    'moderate'   44-5, 65, 117, 118
repeater   51-2
repetition code   123, 127
ring (algebraic)   152-3, 154, 163
robustness of digital communication   2-3
roll-off   53, 54
Roman alphabet   41, 42, 204, 206
RSA (Rivest-Shamir-Adleman) cipher   223-5

satellite communications   7, 11, 116, 118, 130
S-box   217–18
scalar (algebraic)   161
scalar product   96
secrecy, theoretical   208–10
secret, sharing a   168
sender   203
sha function   34; *see also* delta comb
Shannon's theorem   111–15
signal constellation   64
signal space   94
   dimension of   95
signalling
   narrow-band   6, 60, 115, 200
   orthogonal   107–11
   polar   43, 86–9
signature   204
sinc function   14
single-parity-check code   123, 126, 141
Slepian array   149
'smooth' function   15
SNR (signal-to-noise ratio)   11
source coding, *see* data-compression
source of information   2
space mission   12
spherical code   107, 140
spread-spectrum communications   115–16
standard array   149
stationary condition   78
Stirling's formula   67, 198
stochastic process   77
subfield   156–7
subkey   219
submarines, communication with,   7, 61

suppression of thermal noise   196–7
synchronization of stream cipher   215, 216
syndrome   137–8, 139, 149, 159–60, 174–5, 180

TDM (time-division multiplexing)   50
threshold of hearing   9
training of adaptive equalizer   57
travelling-wave tube   65
trellis code   128–33, 148
trial (in probability theory)   70
triangle inequality   133

UHF (ultra-high frequency) band   7, 9
uncertainty principle   196
unicity distance   209
union bound   70, 71
   tangential   101, 109–10
uniqueness property (in Fourier theory)   18, 27, 30, 31
unit of signal amplitude   9

variance (in probability theory)   72, 73, 76
vector quantization   107
VHF (very-high frequency) band   7, 9
VLF (very-low frequency) band   7, 61

wavelength   6, 7, 8, 185
windowing   25, 28; *see also* convergence factor

Zech's logarithms   156, 157